# SpringerBriefs in Environmental Science

SpringerBriefs in Environmental Science present concise summaries of cutting-edge research and practical applications across a wide spectrum of environmental fields, with fast turnaround time to publication. Featuring compact volumes of 50 to 125 pages, the series covers a range of content from professional to academic. Monographs of new material are considered for the SpringerBriefs in Environmental Science series.

Typical topics might include: a timely report of state-of-the-art analytical techniques, a bridge between new research results, as published in journal articles and a contextual literature review, a snapshot of a hot or emerging topic, an in-depth case study or technical example, a presentation of core concepts that students must understand in order to make independent contributions, best practices or protocols to be followed, a series of short case studies/debates highlighting a specific angle.

SpringerBriefs in Environmental Science allow authors to present their ideas and readers to absorb them with minimal time investment. Both solicited and unsolicited manuscripts are considered for publication.

More information about this series at http://www.springer.com/series/8868

Vivek Shandas · Cynthia Skelhorn ·
Salim Ferwati

# Urban Adaptation to Climate Change

## The Role of Urban Form in Mediating Rising Temperatures

 Springer

Vivek Shandas
Toulan School of Urban Studies and
Planning
Portland State University
Portland, OR, USA

Cynthia Skelhorn
Qatar University
Doha, Qatar

Salim Ferwati
Qatar University
Doha, Qatar

ISSN 2191-5547                    ISSN 2191-5555   (electronic)
SpringerBriefs in Environmental Science
ISBN 978-3-030-26585-4          ISBN 978-3-030-26586-1   (eBook)
https://doi.org/10.1007/978-3-030-26586-1

This Springer imprint is published by the registered company Springer Nature Switzerland AG
The registered company address is: Gewerbestrasse 11, 6330 Cham, Switzerland

# Acknowledgement

The authors are grateful for several organizations and individuals who helped provide feedback, guidance, support, and encouragement for the completion of this book. We acknowledge NPRP grant # NPRP 5-074-5-015 from the Qatar National Research Fund (a member of Qatar Foundation), which helped to provide financial support for hiring the students, post-docs, and other staff, who conducted much of the research herein. Portland State University – specifically the Institute for Sustainable Solutions and the Toulan School of Urban Studies and Planning – was also instrumental in providing financial, logistical, and additional research support for the studies presented in this book. The authors would like to acknowledge Paul Pawlowski, David Sailor, and Greg Acker, all of whom provided insights and guidance for working across continents, disciplines, and the research and practice divide. We also drew deep inspiration from the many students and staff at our respective universities and workplaces who reminded us that future generations will inherit today's infrastructure decisions, and without considering the amplification of heat (and the myriad other climate-induced hazards), we will face fewer options for adapting to a dysfunctional climate system.

# Introduction: Why Urban Adaptation and Why Now?

Vivek Shandas

Upon the writing of this book, organizations, country representatives, scientists, and policymakers are meeting to revisit the "Paris Agreement", a voluntary agreement signed by 184 parties that agree to keep the increase in global average temperatures to well below 2 degrees Celsius above preindustrial levels, and to limit the increase to 1.5 degrees Celsius. Like its predecessor, the Kyoto Protocol, which sets commitment targets that have legal force, the Paris Agreement emphasizes consensus building, allowing countries to voluntarily determine targets. The politically driven agendas of many countries, such as the United States, China, Qatar, and Brazil, have made such international agendas idiosyncratic and subject to changes in political power. As a result, greenhouse gases are on a trajectory to surpass the 2 degree Celsius by the end of the century, if not sooner, and scientific papers[1] suggest that society has less than a 5% chance to achieve the targets set forth in the Paris Agreement. By all measures, the ongoing negotiations, whether maintaining momentum for the current Paris Accord or future events, seem all but impossible.

This book aims to understand the intersection of two central concepts that emerge to define the future of cities: *adaptation and livability*. While we examine the concept of livability in the next chapter, here, we argue that the well-documented changes in our climate system, and the breakdown of global agreement on capping emissions, will require monumental preparation for the unprecedented changes already occurring to the systems that define the modern society. While literature on the grave impacts of climate change is mounting, few studies offer empirical evidence about the role that our built environmental plays in amplifying planetary warming. By conducting empirical assessment of one of the cities on the brink of extreme temperature inhabitability—Doha, Qatar—we examine the relationship between adaptation and livability by addressing several questions that help to define the emerging field of urban climate adaptation studies. Some of these questions include, how do increasing temperatures interact with land use change to amplify experiences of urban heat? To what extent can interventions

---

[1]Raftery, AE, A Zimmer, DMW Frierson, R Startz, and P Liu 2017. Less than 2 °C Warming by 2100 Unlikely, Nature Climate Change Vol. 7: 637–641.

to the current urban form help to temper extreme heat in neighborhoods? How might future changes in land use mediate opportunities for adaptations? Each chapter provides a glimpse into these questions, and we begin by providing a broader frame of the potential challenges facing cities in the face of a warming planet.

## *Towards Adaptation*

Researchers around the world are beginning to grapple with changes in weather, ecosystems, agriculture, disease patterns, and physical infrastructure that are predicted under a scenario of 2 °C warming by 2100. Relying on paleoclimate records, the science boils down to ten climate-relevant impacts, and a few social ones. The most notable for the purposes of this book is the implication that of a 2 °C warming will result in increases in frequency of extreme weather events, such as hurricanes, flooding, heatwaves, drought, large and rapidly spreading wildfires, rising sea levels, and changes to environmental conditions and ecosystems, which are leading to changing conditions for plant and animal life. Additional implications include the destabilizing of social and economic systems, crop failures, and the expansion of diseases for which many communities have little experience. These implications will occur with widely varying frequency, though current climate models suggest increasing intensity and duration of extreme weather events. At the local scale, communities that are mal-adapted to withstand the change in conditions will likely experience the greatest impact, while those that are not preparing at all will face grave consequences from which they may never recover.

We argue that the effects of these extreme climate events are especially acute in urban areas, where society, ecosystems, and technology come together as interconnected systems. Failures to major infrastructure systems such as transportation (e.g., airplanes, roads, railroads), energy (e.g., transmission lines, power generation, overload demand), water (e.g., pipe integrity, supply shortages, distribution capacity), and others can generate a cascading series of impacts that ultimately befall the most vulnerable communities and ecosystems (Kim et al. 2016). In particular, extreme climate-induced events reduce the capacity to accommodate acute and additional pressures for which many urban systems were not designed. Impacts on community health and well-being alone, for example, during a heatwave take the lives of more people than all other natural disasters combined (Klienenberg 2004). Moreover, communities that have less access to and control of resources (e.g., financial, land, infrastructure, social) are more susceptible to extreme events than others urban dwellers, in part, due to historical environmental injustices that hamstring preventative actions.

One sobering interpretation of unabated increases in temperatures, which the UN estimates upwards of four degrees Celsius by the end of this century, will cause damage to urban infrastructure valued at over $600 trillion[2]—a value that is double the total global wealth today. In fact, as soon as 2050, people in places like Doha and Delhi will face a lethal risk of setting foot outside in the summer. These temperature conditions may create cascading effects in the form of climate refugees—those who are moving out of uninhabitability places. The UN's conservative estimate for the number of climate refugees that could be produced by 2050 is upwards of 100 million people. These are implications that we are beginning to see already in many parts of the world, though we will also encounter other cascading effects currently unknown.

Urban climate adaption may be a direct means for improving the livability of regions undergoing major climate stress. Yet to understand the what systems may be most at risk and to what climate stressor, we will first need to understand, empirically, the relationship between the current form of our cities and the extent to which we can modify and temper potential impacts. One way to advance an adaptation agenda is to provide evidence about the current distribution of climate-induced events, explore options for mitigating impacts, and project future scenarios for meeting livability goals.

To that end, and to illustrate a systematic approach to addressing climate-induced hazards, this book seeks to identify the distribution of one climate stressor—urban heat—and examine the opportunities for adapting the built environment for increasing temperatures. Urban heat emerges as a result of trapping solar radiation, and amplifying it through the built environment. While an ever-expanding literature points to the known and novel effects of the urban heat, including the acceleration of several natural processes (e.g., evaporation, net primary productivity, species migration, etc.), our intention is to assess the characteristics of the features we can directly control through land-use policy and programs. We posit that by assessing the capcity of the built environment to amiliorate high temperatures, we can reduce exposure to extreme urban heat, thereby reducing fatalities and improving livability.

We focus on urban heat because it is a "silent killer". Outdoor daytime air temperatures in many cities test the human body's ability to tolerate being outdoors throughout much of the year. The rapid rise in urban populations, combined with increasing frequency of extreme heat events, increases the likelihood of communities suffering from respiratory illnesses, heatstroke, and cardiac failure (Luber and McGeehin 2008; Reid et al. 2009). Heat stress is, in fact, one of the leading weather-related causes of death in many parts of the globe (Knowlton et al. 2007; Balbus and Malina 2009). Therefore, finding ways to reduce the intensity of urban heat stress poses an important challenge to public health, tourism, and the livability of cities in general.

Our focus is on using the land-use planning system to understand and ameliorate potential exposure to extreme heat. Characteristics of the built environment generate a phenomenon generally known as the urban heat island (UHI) (Oke 1995;

---

[2]*United Nations, 2018. The Summary for Policymakers of the Special Report on Global Warming of 1.5 °C (SR15). Intergovernmental Panel on Climate Change, United Nations.*

Voogt and Oke 2003). The UHI occurs where measurable differences in both air and surface temperatures are found between an urban area and surrounding rural areas (Oke 1969; Landsberg 1981). A range of factors, including the city size, as measured by population, the increased density of human-made structures and surface materials that are drier than their surroundings and radiate sensible heat, and anthropogenic sources of heat, such as waste heat from vehicles and buildings, are known to contribute to the UHI (Oke 1982; Golden 2004; Levermore et al. 2015). The regional-scale description of urban heat complements an emerging body of evidence that describes differences in temperatures within metropolitan areas.

This regional-scale description of UHIs complements an emerging body of evidence that describes local thermal anomalies (LTAs)—those areas within a city that are relatively hotter than other areas. In fact, a growing body of literature emphasizes the role of the built environment as a mitigation strategy for UHI (Younger et al. 2008; Santamouris, Synnefa and Karlessi 2011; Aflaki et al. 2017; Gunawardena, Wells and Kershaw 2017; Santamouris et al. 2017). Commonly proposed interventions include tree planting, use of green roofs, and an overall increase in green spaces (Oliveira, Andrade and Vaz 2011; Santamouris 2014; Upreti, Wang and Yang 2017), as well as lightening roads, roofs, and buildings to increase albedo (Radhi et al. 2017; Kyriakodis and Santamouris 2018). Further evidence suggests that desert cities, unlike cities in temperate zones, often show a UHI effect, inverting the urban heat island phenomenon, with the result that specific urban areas appear colder than suburban areas during the daytime (Lazzarini, Marpu and Ghedira 2013; Rasul, Balzter and Smith 2015). As a result, urban development patterns have the potential to reduce temperatures and increase accessibility to the outdoor environment through modifications to the built environment. Previous research in the arid desert cities of Phoenix, Dubai, and others suggests that a combination of vegetation, the presence of water, and landscape design all affect the thermal comfort of human inhabitants (Brazel et al. 2007; Nassar, Alan Blackburn and Duncan Whyatt 2014).

Adapting urban environments to address LTAs requires an understanding of the role of land-use planning in ameliorating urban heat, though the differences across places can often hinder rapid deployment of relevant information. This book is admittedly an early contribution to understanding urban climate adaption efforts, though the empirical analysis contained herein offers glimpses into the possibilities for sustaining urban places, and improving the chances of making them livable during a rapid acceleration of global temperatures. We believe that anybody with a general interest in climate and cities will find in the following seven chapters, a promising set of practices that help to prepare regions for a warming planet. The ideal audience would consist of an individual or groups who, either through their own or external interests, recognize the severity and formidable challenge of transforming cities into climate-adaptive landscapes. This book is not a political statement, nor does it aim to engage only those who "believe" in climate change; rather, it offers a pragmatic and empirical assessment about the potential consequences of greater amounts of greenhouse gases in the atmosphere, and the approaches for enabling cities and their communities, infrastructure, and ecosystems to cope. Professionals—current and emerging—working in the fields of city and regional planning, public health, architecture, landscape architecture, natural resource management, and community development will find familiarity in these concepts.

We begin by describing the relationship between urban heat and livability, underscoring historical and contemporary interpretations. We then provide an analytical description of our case study city—Doha, Qatar—as it grew from a small pearling village on the Persian Gulf to a global city, whose infrastructure has grown at a blistering pace (Chap. 2). By examining the entire region, we are then able to examine specific areas of the city and how they vary in terms of exposure to urban heat (Chap. 3). These neighborhood-scale analyses are where we find the profound implications between the built environment and temperature differences (Chap. 4). By selecting specific neighborhoods of Doha, we are then able to examine how alternative urban designs can mediate temperatures and reduce exposure to extreme urban heat. We also examine how future projects of urban development might vary as a result of instituting an urban growth boundary (Chap. 6). We conclude by describing potential changes in the built environment, speculating on changes in exposure to urban heat, and identifying planning mechanisms that can improve the adaptability and livaiblity of increasingly uninhabitable cities (Chap. 7). We believe that this empirical assessment will be instrumental in showcasing and modeling approaches that other regions can employ for understanding the relationship between climate stressors and opportunities for urban adaptation.

# References

Aflaki A, Mirnezhad M, Ghaffarianhoseini A, Ghaffarianhoseini A, Omrany H, Wang ZH, Akbari H (2017) Urban heat island mitigation strategies: a state-of-the-art review on Kuala Lumpur, Singapore and Hong Kong. Cities. Pergamon 62:131–145. https://doi.org/10.1016/J.CITIES.2016.09.003

Balbus JM, Malina CM (2009) Identifying vulnerable subpopulations for climate change health effects in the United States. J Occup Environ Med 51(1):33–37. https://doi.org/10.1097/JOM.0b013e318193e12e

Brazel A, Gober P, Lee SJ, Clarke-Grossman S, Zehnder J, Hedquist B, Comparri E (2007) Determinants of changes in the regional urban heat island in metropolitan Pheonix (Arizone, USA) between 1990 and 2004. Clim Res 33:12. https://doi.org/10.3354/cr033171

Golden JS (2004) The built environment induced urban heat island effect in rapidly urbanizing arid regions—a sustainable urban engineering complexity. Environ Sci Taylor & Francis Group 1(4):321–349. https://doi.org/10.1080/15693430412331291698

Gunawardena KR, Wells MJ, Kershaw T (2017) Utilising green and bluespace to mitigate urban heat island intensity. Sci Total Environ. The Author(s). 15(584–585):1040–1055. https://doi.org/10.1016/j.scitotenv.2017.01.158

Knowlton K, Lynn B, Goldberg RA, Rosenzweig C, Hogrefe C, Rosenthal JK, Kinney PL (2007) Projecting heat-related mortality impacts under a changing climate in the New York City region. Am J Publ Health 97(11):2028–2034. https://doi.org/10.2105/AJPH.2006.102947

Kyriakodis G-E, Santamouris M (2018) Using reflective pavements to mitigate urban heat island in warm climates—results from a large scale urban mitigation project. Urban Clim. Elsevier 24:326–339. https://doi.org/10.1016/J.UCLIM.2017.02.002

Landsberg HE (1981) The urban climate, vol 28 (International Geophysics), 1st edn. Academic Press, New York

Lazzarini M, Marpu PR, Ghedira, H (2013) Temperature-land cover interactions: the inversion of urban heat island phenomenon in desert city areas. Remote Sens Environ 130:136–152. https://doi.org/10.1016/j.rse.2012.11.007

Levermore GJ, Parkinson JB, Laycock PJ, Lindley S (2015) The urban heat island in Manchester 1996–2011. Build Serv Eng Res Technol 36(3):343–356. https://doi.org/10.1177/0143624414549388

Luber G, McGeehin M (2008) Climate change and extreme heat events. Am J Publ Health 35:429–435

Nassar AK, Alan Blackburn G, Duncan Whyatt J (2014) Developing the desert: the pace and process of urban growth in Dubai. Comput Environ Urban Syst 45:50–62. https://doi.org/10.1016/j.compenvurbsys.2014.02.005

Oke TR (1969) Towards a more rational understanding of the urban heat island. Climatol Bull (McGill University) 5:1–20

Oke TR (1995) The heat island of the urban boundary layer: characteristics, causes and effects. In: Wind climate in cities. Springer, Dordrecht, The Netherlands, pp 81–107

Oke TR(1982) The energetic basis of the urban heat island. Q J R Meteorol Soc 108(455):1–24

Oliveira S, Andrade H, Vaz T (2011) The cooling effect of green spaces as a contribution to the mitigation of urban heat: a case study in Lisbon. Build Environ 46(11):2186–2194. https://doi.org/10.1016/j.buildenv.2011.04.034

Radhi H, Sharples S, Taleb H, Fahmy M (2017) Will cool roofs improve the thermal performance of our built environment? a study assessing roof systems in Bahrain. Energy Build 135:324–337. https://doi.org/10.1016/j.enbuild.2016.11.048

Rasul A, Balzter H, Smith C (2015) Spatial variation of the daytime surface urban cool island during the dry season in Erbil, Iraqi Kurdistan, from Landsat 8. Urban Clim 14:176–186

Reid CE, O'Neill MS, Gronlund CJ, Brines SJ, Brown DG, Diez-Roux AV, Schwartz J (2009) Mapping community determinants of heat vulnerability. Environ Health Perspect 117:1730

Santamouris M (2014) Cooling the cities—a review of reflective and green roof mitigation technologies to fight heat island and improve comfort in urban environments. Sol Energy. Pergamon 103:682–703. https://doi.org/10.1016/J.SOLENER.2012.07.003

Santamouris M, Ding L, Fiorito F, Oldfield P, Osmond P, Paolini R, Prasad D, Synnefa A (2017) Passive and active cooling for the outdoor built environment—analysis and assessment of the cooling potential of mitigation technologies using performance data from 220 large scale projects. Sol Energy 154:14–33. https://doi.org/10.1016/j.solener.2016.12.006

Santamouris M, Synnefa A, Karlessi T (2011) Using advanced cool materials in the urban built environment to mitigate heat islands and improve thermal comfort conditions. Sol Energy. Pergamon 85(12):3085–3102. https://doi.org/10.1016/J.SOLENER.2010.12.023

Upreti R, Wang ZH, Yang J (2017) Radiative shading effect of urban trees on cooling the regional built environment. Urban Forestry & Urban Greening. Urban & Fischer 26:18–24. https://doi.org/10.1016/J.UFUG.2017.05.008

Voogt JA, Oke TR (2003) Thermal remote sensing of urban climates. Remote Sens Environ 86(3):370–384

Younger M, Morrow-Almeida HR, Vindigni SM, Dannenberg AL (2008) The built environment, climate change, and health: opportunities for co-benefits. Am J Prev Med 35(5):517–526. https://doi.org/10.1016/j.amepre.2008.08.017

# Contents

# Chapter 1
# Urban Heat and Livability

**Vivek Shandas**

**Abstract**  The rapidly changing climate is a crisis that, at its core, challenges many of the systems that support human habitation in cities, where the majority of people now live. The local effects of our destabilizing climate are most profound through extreme events, and heat waves are the most notorious for killing more people than all other natural disasters combined. In places like Doha, Qatar, where ambient temperatures can reach upward of 50 °C, will, by necessity, be at the forefront of adaptation strategies to improve livability through mediating extreme heat. Even though the region has long considered extreme heat in planning urban development, increasing temperatures require further reformulation of planning and development systems to accommodate new questions relevant to livability. This chapter provides a description of the relationship between rapidly growing urban regions, such as Doha, and the relationship between the built environment and human livability.

**Keywords**  Built environmental · Extreme urban heat · Planning systems

One of the more profound and insidious—yet largely misunderstood—disasters to affect humans and the built environment is urban heat. Exposure to heat claims more lives than all other natural disasters combined, yet fatalities to humans and impacts on infrastructure are largely preventable. While the radiative heating of urban infrastructure from heat waves directly impacts communities, the latter are also indirectly affected through the drying of surrounding lands and an increasing likelihood of devastating wildfire smoke. Arguably, communities that are less prepared will face greater impacts from heat waves, and those that have the least access to cooling resources will encounter the greatest fatalities. Indeed, as the demographic profiles of countries change, strategies aimed at improving resilience to natural disasters will need to recognize how different cultures and socioeconomic populations perceive, design, and are affected by infrastructure. Ameliorative strategies will also need to consider the specific geographic contexts and the bioclimatic changes that climate change will further exacerbate (Voelkel et al. 2018).

© The Author(s), under exclusive license to Springer Nature Switzerland AG 2020
V. Shandas et al., *Urban Adaptation to Climate Change*,
SpringerBriefs in Environmental Science,
https://doi.org/10.1007/978-3-030-26586-1_1

Expanding urban areas are now home to 50% of the world's population and projected to hold 70% by 2050. Across the rapidly industrializing regions, people are moving into cities at unprecedented rates. The literature on global migration suggests that urban in-migration occurs due to "push" and "pull" factors. Push factors consist of factors that make the current nonurban living difficult to access opportunities for economic prosperity, education, health care, and other essential services. The capacity of cities to support growing numbers of people while providing these services points to the emergence of megacities, whose boundaries aim to house over ten million people. While the creation of these megacities is largely accidental due to massive increases in population over time, some countries like China are aspiring to deliberately create "super cities" consisting of urban regions that cover hundreds of square kilometers, contain high-speed train corridors connecting central hubs, and provide essential services for upward of 25 million residents.

The intersection of massive migration into cities and the acute pressures from a changing climate brings with it a need to better understand how different cities are responding. While the urban ecology literature points to a relationship between rapidly growing urban centers and the concomitant pressures on both adjacent and more distant ecosystems, policy-makers and practitioners often use untested assumptions and incomplete or biased information without understanding the complicated interactions and feedback loops between humans, their settlement patterns, and ecosystems. We would expect cities to have well-established responses to safeguard communities from climate-induced stressors; however, an emerging body of evidence—in the form of empirical assessments, literature evaluations, and germane conferences—suggests that very few cities are adequately prepared to address short-term response or long-run resilience to naturally occurring events such as earthquakes, heat waves, floods, or fires (Solecki et al. 2011). Attempts to systematically reduce exposure to climate impacts within cities are further stymied by several factors, including bureaucratic fragmentation (Carruthers 2003), perceptions of safety, and concerns that immediate needs take priority over plausible future events.

While these observations are not novel or groundbreaking, they identify that robust understanding and planning is still needed to prepare for the changes in the climate system, including a focus on cities, where the majority of humans now live. In fact, reducing the exposure of humans to heat is quickly becoming one of today's grand challenges. Exposure to urban heat is mediated in a myriad of ways, though cities and their built environments are epicenters for amplification of impacts from the urban heat. The exposure of human populations to urban heat is largely mediated through the buildings that serve as the frontline of defense from hot and polluted ambient environments. Buildings comprise the majority of the infrastructure in our cities and serve as the frontline of defense against human exposure to extreme events. Yet, remarkably few studies are examining the capacity of buildings and their neighborhoods to withstand shocks from extreme heat; as a result, the majority of U.S. buildings are woefully underprepared for, and disproportionately affected by, extreme weather events (Cutter 2016). Fundamentally, the built environment uses energy and materials to support the health and comfort of occupants. In doing so, buildings in the urban areas of Doha consume nearly 70% of primary energy produced, compared to ~40% in the United States (Brookings 2015). Since US and Qatar energy is primarily of fossil origin—either through crude oil or natural gas—building energy use contributes to climate forcing that exacerbates heat waves.

The rapid changes occurring in Qatar and throughout the Gulf region could result in serious environmental and social problems and accelerate global environmental change unless society develops an understanding of the drivers of urbanization, its impacts, and opportunities for mitigation. Though the immediate impacts of climate-induced events such as hurricanes, heat waves, flooding, and fires are abundantly available in everyday news, we approach the topic of adaptation planning by looking further ahead and focusing specifically on the built environment—the infrastructure, design, and qualities of neighborhoods that will make them more or less livable.

We use the concept of livability to identify factors in the built environment that can help to facilitate our capacity to spend time comfortably outdoors. Throughout the book, we reference traditional, alternative, and regenerative approaches to urban design that can help society better cope with the accelerating changes in our climate system. Additionally, our focus on urban heat offers tractable direction for those systems of neighborhood design that require further scrutiny.

Places that are already hot will endure wet-bulb thresholds that prevent people from spending time outside. In fact, due to its hot climate from late mid-April through November and the expanding infrastructure, recent studies confirm that urban areas in the Middle East and North Africa are arguably going to see the earliest exposure to wet-bulb temperatures, in which the ambient outside temperatures can no longer evaporate human sweat. The wet-bulb threshold is of keen interest to epidemiologists and medical professionals because human exposure to these temperatures is almost certainly fatal. Generally speaking, at an outdoor wet-bulb temperature of 35 °C, the human body experiences the wet-bulb threshold and heat exhaustion and respiratory illness follows shortly thereafter.[1] Current projections indicate that by the middle of twenty-first century, areas of the Middle East and North Africa (MENA) will have temperatures that will not drop below 30 °C at night and potentially hit 46 °C during the day. During the warm periods of the year, midday temperatures could reach over 50 degrees. Humans can adapt to high temperatures as long as humidity remains low and hydration and sweating can occur. As the planet warms, and more moisture is available in the air column, we can estimate that even temperate areas will have an increase in humidity.

Interestingly, while regional temperatures may reach upward of 46 °C, our studies suggest that, during the midday, temperatures can vary by upward of 12 °C within the same hour and in one urban area. Consistent among the studies of urban heat is the fact that the materials, landscaping, moisture content, configuration of buildings, and other design features can help to mitigate the exposure to extreme heat. If urban designs can provide some relief, we reason that adaptation strategies for coping with extreme heat (and concomitant climate-induced stressors) require further atten-

---

[1] Sherwood, S. C. and Huber, M. (2010) 'An adaptability limit to climate change due to heat stress', Proceedings of the National Academy of Sciences, 107(21), p. 9552 LP-9555. https://doi.org/10.1073/pnas.0913352107.

tion. Furthermore, with rapid urbanization occurring throughout the MENA region, perhaps a closer examination of this region may provide evidence for improving livability in other urban areas as the planet warms.

What will cities of the future need to look like to make them livable, sustainable, and able to withstand imminent climate pressures? How will building designs vary as a result of extreme urban heat? How can alternative approaches enable communities to withstand, indeed adapt to, the ever-increasing urban temperatures? Although a robust treatment of any of these questions is not possible in a single book, we identify promising practices that can better equip our cities for a warming planet. We use the concept of livability as an intersectional concept that brings together human health and well-being, accessibility to essential goods and services, and ambient environmental conditions that support communal gathering in outdoor spaces. Together these concepts help to reveal the underlying factors that are under the control of human decision-making and city planning, and which can support safe human habitation. We imagine that creating more livable communities—ones that enable humans from diverse socioeconomic, ethnic/racial backgrounds to live harmoniously with the planet—is a direction that crosses disciplinary and political persuasions, to which we can all strive. The climate crisis will require no less.

# References

Brookings (2015) Navigating uncertainty: Qatar's response to the global gas boom. Brookings Institution

Carruthers J (2003) Growth at the fringe: the influence of political fragmentation in United States metropolitan areas. Reg Sci 82(4):475–499

Cutter SL (2016) The landscape of disaster resilience indicators in the USA. Nat Hazards 80(2):741–758

Solecki, William & Leichenko, Robin & O'Brien, Karen (2011) Climate change adaptation strategies and disaster risk reduction in cities: Connections, contentions, and synergies. Curr Opin Environ Sustain 3:135–141

Voelkel J, Hellman D, Sakuma R, Shandas V (2018) Assessing vulnerability to urban heat: a study of disproportionate heat exposure and access to refuge by socio-demographic status in Portland, Oregon. Int J Environ Research and Public Health 15(4):640

# Chapter 2
# Urban Transformations, Past and Present

**Vivek Shandas, Cynthia Skelhorn and Salim Ferwati**

**Abstract** Changes to the built environment are a result of myriad decisions over the decades (and sometimes centuries) by policy-makers, architects, planners, and landscape designers. Our cities reflect historic decisions that have crafted the built environment and our capacities to transform landscapes to address emerging and acceptable threats to our health and safety. This chapter traces the history of cities, identifies examples of current movements, and begins to define a twenty-first-century concept of livability. While we focus on the interaction of the built environment with governance systems, we highlight Doha's transformation, over time, to argue that the study of this city is important, not only because it represents the trajectory and ambitions for thousands of others but also because contains many of the characteristics that make it ideal to study urban adaptation. Results from climate models suggest that Doha will be uninhabitable, further making it a poster child for taking urban measures for safeguarding its rapidly growing population.

**Keywords** Cities in history · Suburbanization · Livability · Adaptation

Some places in the world are experiencing urbanization rates that are unprecedented. The relationship between the rate of growth and livability is complex, and single indices that describe sprawl or sustainability do not provide the depth of analysis needed for addressing the climate crisis. In the West, the industrial revolution of the 1800s created harsh urban living conditions, which led to the creation of the urban planning system. The early 1900s consisted of a series of urban reform movements that laid the foundation for increasing livability, which at the time meant public health and safety. Lewis Mumford depicts this era in his classic Cities in History (1962) book as one consisting of "stench and disgust," to which many ambitious planners devoted their careers. Individuals like Le Corbusier, Sir Patrick Geddes, and Mumford himself saw the emergence of the industrial revolution as a call for city planning; their focus was to improve urban livability. Movements such as the City Beautiful, The Garden City, and the Radiant City provided the basis for organizing

© The Author(s), under exclusive license to Springer Nature Switzerland AG 2020    5
V. Shandas et al., *Urban Adaptation to Climate Change*,
SpringerBriefs in Environmental Science,
https://doi.org/10.1007/978-3-030-26586-1_2

the urban landscapes to accommodate an ever-increasing population, while ensuring public health and safety, and social cohesion.

The decades following these early urban reform movements laid the theoretical and practical applications of current urban planning systems. Land-use laws, zoning, and regulating private enterprise provided the backbone to ensure the sustainable availability of common goods and resources. These mechanisms were continually challenged though arguably the decades following the 1980s created a profound change to the capacity for cities to regulate development, and hence their livability. The massive deregulation of industries across North American and Europe and the creation of free trade agreements across the 1980s opened a global race for capital that continues to the present. Cities serve as the epicenter for the globalization of markets because they provide the technological and social infrastructure for enabling communication and legalization systems. As a result, modern interpretations of livability sought an unequivocal focus on increasing income. By increasing income, proponents claimed, cities will thrive and become more competitive in the global pursuit of capital. Through the 1980s and 1990s, those regions that emerged victorious in attracting capital were deemed "global cities," since they drove international finance, cultural industries, and massive populations. Four cities—Tokyo, Paris, London, and New York—directed the production and consumption of goods and services. Tokyo, for example, with its 30 million inhabitants, commanded over 90% of Japan's GDP, and had nearly 100% employment. New York, Paris, and London had GDP levels larger than most countries of the world.

While the importance of rising income continues to be a dominant narrative within the international development and economics literature, several fissures began to reveal its limitations when examining livability. One that rose to prominence is the loss of community and the isolation of individuals from the rest of society. While expanding employment, income, and gross domestic product may support national agendas for competition, notions of social cohesion, which early 1900s urban scholars aspired to achieve, fell by the wayside. Prominent figures, such as Jane Jacobs, Glenn Loury, Robert Putnam, Elinor Ostrom, and others (Membiela-Pollán and Pena López 2017), challenged the "income above all" narrative and started to generate modern theories about the importance of community and sustainability. These critiques questioned the fundamental basis for a never-ending growth model that isolated individuals and gobbled up more and more of the finite planetary resources. Measurement systems, such as bridging and bonding capital emerged, and cities, businesses, and organizations started to develop policies that aimed to improve collective identity. A series of notable publications (e.g., Kunstler's The Geography of Nowhere, Putnam and Feldstein's Better Together, etc.) also challenged the suburban single-family development model using empirical assessments to describe increases in the isolation of families, dependence on automobiles, and obesity and other health crises. Cities such as Portland, Oregon (USA) and Curitiba, Brazil began experimenting with growth management through boundaries and integrated transportation systems. Other cities with ambitions to improve livability soon followed with experiments of their own.

Fueled by expanding global capital, the 1990s and 2000s were instrumental in exporting the US and Western Europe's models of "development" to Asia, Latin America, and Eastern Europe. The automobile, which was the basis for much of the urban form in the US, was making its way into other parts of the world, and soon to follow were familiar urban development programs. Often through structural adjustment loans from multilateral lending institutions (e.g., World Bank, IMF, Asia Development Bank, etc.), cities throughout Asia, Latin America, and Europe began creating suburbs, complete with large malls, office parks, and private gated communities. Even with their critiques, urban planners in the rapidly urbanizing regions of the world quickly began applying modeling of suburban development, making way for a primary focus on the automobile. The horse had left the stable, and the race for rapid industrialization was underway.

While suburbanization, automobile dependence, and increasing dependence on fossils fuels are adopted around the world—often by politicians who are eager to develop cities that can compete for global capital—critiques to rapid and unsustainable urban development are increasingly finding unusual suspects. In fact, since 2010, several concepts of neighborhood development are beginning to challenge the conventional suburban development models that the US exported throughout the world. The concept of an EcoDistrict emerged out of Portland, Oregon in 2009 with a publication that outlined the opportunity and challenge facing neighborhood-scale urban design. Seltzer et al. (2010) provide the basis for integrating concepts of "human scale" urban form with neighborhood-scale governance systems that emphasize sustainability, livability, and social cohesion all at the same time. These concepts have expanded further with a recent publication of Emily Talen's book, Neighborhood (2018), in which she describes the history of the concept and provides examples for advancing meaningful involvement of communities in planning decisions.

While the movement to improve livability through localizing planning decisions to district and neighborhood scales may be gaining attention, needed are descriptions about livability in an era of heightened climate-induced disruptions. Existing studies about natural hazards and urban livability across several disciplines, and social scientists have identified the ways in which engaging communities in decision-making allows for greater adaptive capacity, and how a disruption of social cohesion and/or isolation reduces people's adaptive capacity, making them less capable of enduring environmental stress (Adger et al. 2003). While community engagement may enhance the adaptive capacity for certain individuals, insufficient knowledge or information may negatively affect society's adaptive capacity and, as a result, may stand in the way of people making more permanent adjustments in response to the occurrence, or threat, of longer term environmental change (Brooks 2003). Specialists in particular areas of adaptation (e.g., agriculture or coastal zone management) have identified particular policies, such as enhanced communication of environmental information or the development of insurance networks that can assist in responding to changes in environmental conditions (Freeman and Kunreuther 2002). Likewise, scholars of international relations and international institutions have identified other drivers necessary for effective adaptation planning such as funding and incentive-building mechanisms (Klein et al. 2002).

Although studies of urban livability and climate change recognize that human decision-making and social capacity is cultural and context-specific, emergent are many themes that offer generalizable concepts and can directly apply to urban development. For example, Klein et al. (2005) provide evidence that the coupling of spatial and temporal scales occurs through the direct intervention of governance systems, and the process by which human communities are engaged in the decision-making process. Moreover, during times of rapid urban growth, the information systems, networks, and institutional systems that coordinate these dimensions of adaptation can become fragile and unhinged, potentially leading to catastrophic social and environmental disasters (Freeman and Kunreuther 2002). These studies also suggest a need for better understanding the intersection of climate-induced stressors, urban growth patterns, and effectiveness of adaptation efforts.

The complexity inherent in understanding opportunities for adapting cities to a changing climate requires that they be studied with specific attention to the element of human intent and the nature of information flows. Several variables have been identified as distinct to urban development and not clearly understood in a linked socio-ecological system, including governance and social institutions; the availability of social and human capital; and the capacity for humans to learn from experience. Lebel et al. (2006) characterize attributes of governance that can be important analytically, including participation, representation, deliberation, accountability, empowerment, social justice, and organizational features such as multilayered governance structures and polycentric development patterns. They and others argue that the presence or absence of these features affects the capacity to manage resilience (Carpenter et al. 2009). The conditions for and the extent to which information flows in a decision-making process may be a mechanism for understanding local and regional governance structures and the role of human intent in the coupling of social and ecological systems.

## 2.1  Doha's Transformation

The city of Doha is the capital of the State of Qatar and is located on the Persian Gulf (Fig. 2.1). The primary industry in the city was pearl trading through the 1920s and then collapsed due to the invention of pearl harvesting techniques in Japan. The population peaked at about 27,000 inhabitants in the 1920s (Wirth 1988) and consisted of traditional Islamic desert settlement patterns, as described by Besim Hakim in his book Arab-Islamic Cities: Building and Planning Principles (1986). These patterns consisted of a vernacular road network of cul-de-sacs, which enhanced the privacy of neighborhoods, known as ferej, and the market, located close to the port and constituting the central public realm. After national independence, Doha entered a phase in the 1970s and 80s when the discovery of oil fueled a physical growth of the city's infrastructure. While the old city centers were replaced by commercial buildings and apartment blocks for foreign labor, low-rise housing areas rapidly extended the urban periphery (Al Hathloul 1996).

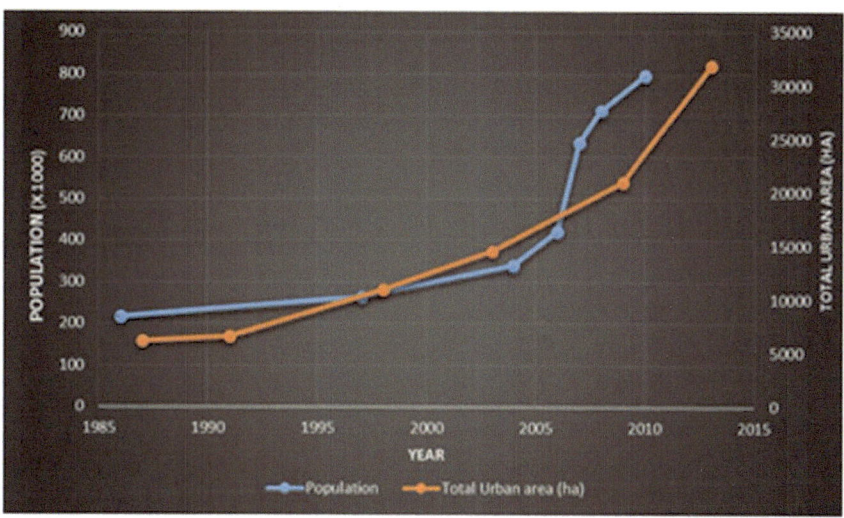

**Fig. 2.1** Doha: population and urban area (*Source* Qatar Information Exchange; Shandas et al. 2014)

At the end of the twentieth century, the Emirate of Dubai introduced a new model of urbanism by implementing growth-oriented development strategies in order to diversify the economy. The subsequent liberalization of local real estate markets led to a vast construction boom and a new chapter of urban development in the Gulf (Schmid 2009). Most cities became hosts to various megaprojects, which were usually launched by newly founded holdings whose main shareholders are usually public institutions due to the fact that most unbuilt land is considered to be the property of the state and thus under the authority of rulers (Wiedmann et al. 2014). New forms of development such as reclaimed islands and high-rise agglomerations emerged and transformed the previous urban morphologies. The new focus on expanding real estate markets as a diversification strategy led to the decentralization of urban governance and the common practice of case-by-case decision-making (ibid.). This rapid growth of Doha development patterns and the immediate availability of satellite imagery offer timely and effective means for understanding the areas where sustainable urban forms are emerging.

Today, Doha is the most densely populated city in Qatar with approximately 40% out of a total population of 2.4 million in 2015 (Qatar Ministry of Development and Planning Statistics 2015). The rates of growth in Doha are particularly acute and rapid during the last 15 years, which can be traced by the relationship between population and infrastructure growth (Fig. 2.1). The increase of urban development that is tied with the increase of population results in expanding impervious areas that can trap heat creating urban heat islands (UHIs).

Many characteristics make Doha ideal for studying climate adaptation planning. Due to its hot climate from late mid-April through November and the expanding infrastructure, residents experience a high degree of thermal discomfort during this part of the year. For example, summer outdoor temperatures frequently exceed 40 °C, whereas humans are most comfortable during summer with ambient temperatures ranging from 23 to 27 °C (Epstein and Moran 2006; Ferwati et al. 2016). These extreme temperatures are noted in the Economist Intelligence Unit's (EIU) Survey of Livability for the Gulf Region. Using benchmarking perceptions of development levels, the EIU quantifies the challenges that any given location might present to an individual's lifestyle and allows for direct comparison between locations. While Doha is not included directly, nearby Manama, Bahrain, and Dubai, UAE are watched regularly and in 2017 were ranked #92 and #75, respectively. For every city, the EIU assigns a rating of relative comfort for 30 qualitative and quantitative factors organized in five general categories: Stability, health care, culture and environment, education, and infrastructure. Each factor is rated as acceptable, tolerable, uncomfortable, undesirable, or intolerable.

Of interest to this study are five specific factors in the category of environment— consisting of humidity/temperature rating and discomfort of climate to travelers— and in the category of infrastructure—consisting of quality of road network, quality of public transport, and availability of good quality housing. Among the environmental factors, the Doha region ranks among the most inhospitable areas of the world, and as such, offers a compelling case for further study about potential adaptation efforts. Collectively, these factors represent and contribute important concerns in defining a larger understanding of livability. Under a business-as-usual scenario, some scientists claim that many areas of the Gulf region are likely to become uninhabitable due to intolerable rises in the wet-bulb temperature (Pal and Eltahir 2016). What can be done about increasing temperatures, and how might the region help others learn how to adapt to a warming planet? This and similar questions will provide the direction for the coming chapters to provide the planning and regional context for better preparing cities.

# References

Adger WN, Huq S, Brown K, Conway D, Hulme M (2003) Adaptation to climate change in the developing world. Prog Dev Stud 3(3):179–195
Al Hathloul S (1996) The Arab-Muslim city: Tradition, continuity and change in the physical environment. Dar Al Sahan, Riyadh
Brooks N (2003) Vulnerability, risk and adaptation: a conceptual framework. In: Tyndall centre for climate change research. Tyndall Centre for Climate Change Research, University of East Anglia, Norwich, 38(November), p 20
Carpenter SR, Mooney HA, Agard J, Capistrano D, Defries RS, Diaz S et al. (2009) Science for managing ecosystem services: beyond the millennium ecosystem assessment. In: Proceedings of the national academy of sciences, 106, pp 1305–1312
Epstein Y, Moran DS (2006) Thermal comfort and the heat stress indices. Ind Health (Japan) 44(3):388–398
Ferwati MS, Shandas V, Sailor DJ, Pawlowski PR, Makido Y, Shawish A (2016) Improving livability in Doha: the role of neighborhood microclimates, land use, and materials in rapidly urbanizing regions. In: ARC '16. Qatar Foundation, Doha, Qatar

Freeman PK, Kunreuther H (2002) Environmental risk management for developing countries. Geneva Pap Risk Insur Issues Pract 27(2):196–214

Klein RJT, Nicholls RJ, Thomalla F (2002) The resilience of coastal megacities to weather-related hazards. In: Kreimer A, Arnold M, Carlin A (eds) Building safer cities: the future of disaster risk. The World Bank Disaster Management Facility, Washington, D.C., pp 101–120

Klein RJT, Schipper ELF, Dessai S (2005) Integrating mitigation and adaptation into climate and development policy: three research questions. Environ Sci Policy 8:579–588

Lebel L, Anderies JM, Campbell B, Folke C, Hatfield-Dodds S, Hughes TP, Wilson J (2006) Governance and the capacity to manage resilience in regional social-ecological systems. Ecol Soc 11(1), part 19. https://doi.org/10.5751/es-01606-110119

Membiela-Pollán M, Pena López J (2017) Clarifying the concept of social capital through its three perspectives: individualistic, communitarian and macro-social. Eur J Gov Econ 6:146–170. https://doi.org/10.17979/ejge.2017.6.2.4327

Mumford L (1962) The city in history: its origins, its transformations, and its prospects. Harcourt, Brace & World, New York

Pal JS, Eltahir EAB (2016) Future temperature in southwest Asia projected to exceed a threshold for human adaptability. Nat Clim Change 6(2):197–200 (Nature Publishing Group). https://doi.org/10.1038/nclimate2833

Schmid H (2009) Economy of fascination. Dubai and Las Vegas as Themed Urban Landscapes, Gebrueder Borntraeger, Stuttgart

Qatar Ministry of Development and Planning Statistics (2015) Statistics, first section: population and social statistics

Seltzer E, Smith T, Cortright J, Bassett EM, Shandas V (2010) Making EcoDistricts concepts & methods for advancing sustainability in neighborhoods. Portland, OR

Talen E (2018) Neighborhood. Oxford University Press

Wiedmann F, Salama A, Mirincheva V (2014) Sustainable urban qualities in the emerging city of Doha. J Urban 1–23

Wirth E (1988) Dubai: Ein modernes städtisches Handels- und Dienstleistungszentrum am Arabisch-Persischen Golf. Selbstverlag der Fränkischen Geographischen Gesellschaft, Erlangen

# Chapter 3
# Rapid Land-Cover Change in Doha

Vivek Shandas, Yasuyo Makido and Salim Ferwati

**Abstract** Amidst the chaotic growth of Asian cities, the expansion of urban infrastructure in the Middle East's Gulf region is arguably outpacing any other region on the planet. Yet, we have a limited understanding of the types of urban form or the extent to which this rapid urbanization is giving rise to sustainable patterns of growth. We ask, what is the pace and character of urban growth in one Middle East city, Doha, Qatar. By using remotely sensed imagery from 1987 to 2013, we examined the pace, quality, and characteristics of urban growth. We further use the results to create a typology of urban growth that integrates historical and spatial dimensions for describing the qualitative aspects of growth and its implications on regional landscapes. Our results suggest that Doha is creating development patterns similar to many Western cities and that planners may need to consider whether the emerging urban form offers opportunities for more sustainable growth in the future.

**Keywords** Rapid urbanization · Development pattern · Remote sensing · Doha, Qatar

The physical growth of cities—a topic that has been studied extensively over several decades—is often used to refer to the physical shape of a given area of development. As of today, empirical evidence about the physical patterns of growth has seen only limited application in non-Western countries. Previous empirical research characterizing urban development patterns suggests a rich and thorough literature, yet still lacking in a global perspective, which primarily focuses on Western regions (Alberti 2010). Examples include the impact of development patterns from transportation corridors and innovation (Ewing and Cervero 2001; Fenkel and Ashkenazi 2008). Other studies also suggest that the process of growth in Western cities impacts prop-

---

Sections of this chapter are from the following document, which is part of the Creative Commons Open Licensing system: Shandas, V, Y Makido, and S Ferwati, 2017. Rapid urban growth and land use patterns in Doha, Qatar: Opportunities for sustainability? European Journal of Sustainable Development Research 1(2): 11.

erty value changes (Glaeser and Kahn 2003; Song and Knaap 2004), density (Parsons Brinckerhoff Quade and Douglas Inc. 1994; Bertaud and Renaud 1997), and network configuration (Loutzenheiser 1997). Tracing land-cover change can provide a means for elucidating development patterns, and the extent to which these patterns give rise to sustainable futures. Those cities that are undergoing rapid landscape transformation—which is the case for many Asian cities (Cho 2005; Shalaby et al. 2012; Li and Peijun 2013; Radhakrishnan 2014)—can be of particular interest to the sustainability discourse since their Western counterparts took several decades to create similar development patterns.

The question of sustainable urban development patterns has generated debate among urban scholars about the economic, environmental, social concerns of growing cities (Cutsinger and Gastler 2006). While sprawling suburban patterns— defined here as auto-dependent low-density development—can create some amenities, including larger living spaces at lower housing costs, and lower crime rates, these low-density developments also cause several problems, including excessive air pollution and oil consumption (Brueckner 2001; Frumkin 2002; Glaeser and Kahn 2003). Moreover, urban sprawl tends to be accompanied by the high costs of building new public infrastructure and commercial developments (Hortas-Rico and Solé-Ollé 2010), which can directly impact the economic sustainability of a region. Indeed, sprawl is one of the most important of several physical patterns of urban growth, but to date, we have limited application of this concept to Middle East cities.

These environmental and social problems, which are often exacerbated by the physical patterns of urban development, point to the need for a better understanding of urban growth patterns outside of American and European contexts. In the past few years, researches in China, India, and Latin America have provided insight into non-Western development patterns (Monkkonen 2008; Li et al. 2009; Bhatta et al. 2010), but these studies are largely untested in the Middle East, where early evidence indicates a combination of Western and Eastern development models (Wiedmann et al. 2012; Rizzo 2013; Nassar et al. 2014).

While a general lack of relevant data still hinders comprehensive assessments of development patterns in the Middle East, recent developments in spatial analysis, computational power, and publicly available satellite data offer a timely and effective means for understanding growth of Middle East cities. For example, assessing urban growth over 50 years in the U.S. can conflate the drivers of growth, such as the rationale for the growth of specific areas and the outcomes of alternative growth patterns. But, when examining physical growth patterns over a shorter period of time during a phase of rapid urban development, we can, arguably, link specific growth processes and reasons for specific patterns of development. We can address such questions as: How does physical growth couple with existing infrastructure to create specific development patterns? How might new forms of development mitigate or exacerbate the environmental and human health impacts of urban growth? The rapid growth occurring in Middle East cities provides an opportunity to answer some of these questions by capturing data over short periods of fast development.

For this chapter, we examined the spatiotemporal coupling of urban growth in one urbanizing region of the Middle East—Doha, Qatar. The city of Doha is the largest city and capital of the state of Qatar and is located on the Persian Gulf. It has one of the fastest growing populations in the Arabic world (Report 2012). While the population of Doha was below 500 thousand just 20 years ago, it is not over 1.2 million and expected to continue rising at unprecedented rates. The rapid rise in population has occurred as a result of discovering natural gas in the mid-twentieth century. The same oil has fueled growth of the physical infrastructure in Doha, and the development of several megaprojects (e.g., 2022 football World Cup, 2006 Asia Games, etc.) has changed the landscape in noticeable ways. Yet, we know of no studies that have documented the land-use patterns since the beginning of the transformation of the physical infrastructure in Doha. As a result, we posit that an understanding of the patterns of growth can shed light into the sustainability of the region, specifically through associating different types of landscape patterns with economic, environmental, and social implications.

Using data from satellite imagery, we analyzed physical changes in the Doha metropolitan region over a recent time period in order to provide a description of urban patterns that are specific to that study region. We asked three research questions: (1) Does rapid urban development in Doha lead to increasing urban sprawl; (2) In what way does rapid urban growth in Doha lead to unique patterns of development; and (3) To what extent does urban sprawl interact with other forms of development to impact the density of urban development? For the purposes of this study, we define urban sprawl as low-density, automobile-dependent development. Answering these questions will provide some of the only regional-scale descriptions of the urban development patterns of Doha. Additionally, we aim to contribute to the use of computer-based analysis of urban areas (e.g., urban geo-informatics) literature by describing the extent to which physical and spatial urban growth interact with temporal scales of development. We begin by describing our methods and relevant results. Then we discuss interpretations of our findings and conclude with a summary of the potential implications of such rapid land-cover change.

## 3.1 Analysis of Land-Cover Change

Our approach to the study of the urban form of Doha relied on satellite sensor imagery and a series of geostatistical analyzes. We acquired 30-m-resolution satellite sensor imagery from the US Geological Survey's Landsat TM, ETM+ and OLI files from the years 1987, 1991, 1998, 2003, 2009, and 2013 (Table 3.1). We selected these images based on the following criteria: (1) cloud cover of less than 5%, (2) summer season (from May to August), and (3) the availability of increments of 4 to 7 years. The adequate imagery at Path/Row:163/42 between 1991 and 1994 were not available at the study site, and therefore we obtained data at Path/Row:162/43 for 1991. We chose to limit our historical period at 1987 because the imagery earlier than that year consists only of Multispectral Scanner (MSS) sensors, which are incompatible to later dates because they have an image resolution of approximately 80 m in four spectral bands ranging from the visible green to the near-infrared (IR) wavelengths.

**Table 3.1** Data used in this study

| Data type | Spatial resolution (m) | Path/Row | Acquisition date YYYY/MM/DD |
|---|---|---|---|
| Landsat5 TM | 30 | 163/42 | 1987/6/14 |
| Landsat4 TM | 30 | 162/43 | 1991/5/25 |
| Landsat5 TM | 30 | 163/42 | 1998/5/11 |
| Landsat7 ETM+ | 30 | 163/42 | 2003/3/14 |
| Landsat7 ETM+ | 30 | 163/42 | 2009/5/1 |
| Landsat7 ETM+ | 30 | 163/42 | 2009/6/2 |
| Landsat7 ETM+ | 30 | 163/42 | 2009/7/4 |
| Landsat7 ETM+ | 30 | 163/42 | 2009/8/5 |
| Landsat8 OLI | 30 | 163/42 | 2013/5/20 |
| QuickBird | 0.6 (Pan-sharpened) | – | 2002, 2003, 2009, 2013 |

Also, the rate of population increase in Qatar was low between 1961 and 1987 (World Bank 2014), and we assume a similar population trend for Doha (MDPS 2014). Thus, in order to avoid the low-resolution imagery and also focus on the rapid change in urban area, we acquired data after 1987. The study area (31.2 km × 37.8 km), which encompassed the Doha metropolitan region, was extracted from each scene of the satellite sensor imagery. Due to the lack of ground truth data in this study site, we obtained QuickBird™ high-resolution imagery and used them for accuracy assessment (Table 3.1). The QuickBird™ satellite was launched in 2001 and provides a very high-resolution imagery, which simultaneously captures 0.61 m panchromatic and 2.44 m multispectral digital imagery.

## 3.2  Data Preparation and Preprocessing

All Landsat bands (visible, near, and shortwave infrared) except the thermal band were used in the classification process. Some images from the year 2009 (Landsat7 ETM+) contain gaps that were reportedly caused by a sensor's hardware failure. For that year, we filled these gaps by mosaicking four images from different months in 2009 into a single composite image. To conduct the accuracy assessment of the Landsat images, the QuickBird imagery was ortho-rectified and pan-sharpened. Orthorectification is the correction of the image, pixel-by-pixel, for topographic distortion, and the result is that every pixel appears to be viewing the earth from directly above (Schowengerdt 1997). Pan-sharpening is a process of combining a lower resolution multispectral imagery with a high-resolution panchromatic imagery to create a high-resolution color image. For this study, Gram-Schmidt Pan-Sharpening (Brower and Laben 2000) was applied to obtain 60 cm multispectral data. All images were georeferenced using first-order polynomial transformation, and overall root mean square

**Table 3.2** Definition of each category

| Classes | Definition |
|---|---|
| Urban | Areas built-up surfaces, including roads, commercial, industrial, pavements, and construction sites |
| Vegetation | Areas containing living plant material, including farms, parks, golf courses, trees, and grasses |
| Soil | Areas containing exposed and non-developed surfaces, including rocks, and coastal and other sands |
| Water | All areas containing water, including reservoirs and the Gulf |

errors were less than half a pixel. All image processing work and accuracy assessments were carried out in ENVI 5.1 (Exelis Visual Information Solutions, Boulder, CO, USA) and ArcGIS 10.2 (ESRI Inc.).

Due to the lack of ancillary data and prior satellite-based research of the study area, we employed a hybrid classification method (Lo and Choi 2004). Each image with six bands was sorted into 40 clusters using the ISODATA unsupervised classification method. Only the homogeneous clusters that coincided with a specific land-use/land-cover type were labeled into one of four categories—Urban, vegetation, soil, and water (Table 3.2). Other clusters that exhibited mixed classes were clipped out from the image for further classification using supervised classification. The maximum likelihood algorithm was applied to the clipped images. The images were initially classified into seven classes: Urban 1, Urban 2, Vegetation, Soil 1, Soil 2, Water 1, and Water 2. Ten to 20 training sites for each class were selected with the aid of a panchromatic band and/or QuickBird imagery. The classified images were resampled to the aforementioned four classes and combined with the ISODATA classified land-use/land-cover map.

## 3.3 Accuracy Assessment

A confusion matrix can provide a basis for describing classification accuracy and for characterizing errors. One of the compelling features of a confusion matrix is its overall accuracy, or percent correctly classified (PCC) (Foody 2002). The PCC is calculated as the ratio of the sum of correctly classified pixels in all classes to the sum of the total number of pixels (Lillesand 2004). The $k$ statistics, or Kappa coefficient, is an indicator of the extent to which the percentage correct values of an error matrix are due to "true" agreement versus "chance" agreement (Lillesand 2004). User and producer accuracy can be obtained from the confusion matrix. The users' accuracy measure indicates the probability that a pixel classified into a given category actually represents that category on the ground; the producer's accuracy measure, which uses a set of training pixels of the given cover type, indicates how well each is classified (Lillesand 2004).

Due to the lack of ground-level data on land use/land cover, we undertook an accuracy assessment using high-resolution satellite sensor imagery only for half of the total images we classified. Unfortunately, the high-resolution imagery for earlier years (1987, 1991, 1998) is unavailable for purchase, and as a result, we only undertook accuracy assessment for years 2003, 2009, and 2013.

We developed a standard confusion matrix by randomly sampling approximately 200 points (50 points for each of the four classes) and overlaying those points on the QuickBird imagery. Each point was visually categorized into one of the four classes by visually determining the dominant land-cover types within 30 × 30-m polygons that coincide with the pixel footprints of Landsat imagery. The categorized data were used as reference data. The matrix was then created to compare the relationship between the reference data and the corresponding classified data (Table 3.3). While the classification accuracy (*k statistic*) for 2013 was 88%, those of 2003 and 2009 were 76% and 70%, respectively. The classification accuracies for 2003 and 2009 do not exceed the minimum acceptable accuracy of 85%, which is adopted by the U.S. Geological Survey (Wilby 2007); however, we note that the level of accepted accuracy can also be an arbitrary measure and that its significance depends on the study area and the purpose of the study (Table 3.4). As we can see from Table 3.5, there is confusion between urban and bare soil class: User's accuracy of bare soil = 63.27%, and producer's accuracy of urban = 65.71%. User's accuracy indicates the probability that a pixel classified into a given category actually represents that category on the ground, and producer's accuracy indicates how well training set pixels of the given cover type are classified (Lillesand 2004). The study area is arid landscape,

**Table 3.3** Land-cover classification accuracy

| Year | Overall accuracy (%) | Kappa coefficient |
|---|---|---|
| 2003 | 78.14 | 0.707 |
| 2009 | 82.35 | 0.764 |
| 2013 | 88.36 | 0.844 |

**Table 3.4** Confusion matrix of accuracy assessment for the land-cover classification for 2009

| 2009 | Class | Reference data | User's accuracy (%) | | | | |
|---|---|---|---|---|---|---|---|
| | | Urban | Vegetation | Bare soil | Water | Total | |
| Classified data | Urban | 46 | 0 | 4 | 0 | 50 | 92.00 |
| | Vegetation | 0 | 45 | 1 | 0 | 46 | 97.83 |
| | Bare soil | 15 | 3 | 31 | 0 | 49 | 63.27 |
| | Water | 9 | 0 | 1 | 32 | 42 | 76.19 |
| | Total | 70 | 48 | 37 | 32 | 187 | |
| | Producer's accuracy % | 65.71 | 93.75 | 83.78 | 100.00 | | |

**Table 3.5** Landscape metrics used in this study

| Landscape metrics | Explanation | Range |
|---|---|---|
| Percentage of Landscape (PLAND) | PLAND equals the percentage the landscape comprised of the corresponding patch type | $0 < \text{PLAND} \leq 100$ |
| Number of Patches (NP) | NP equals the number of patches of the corresponding patch type | $\text{NP} \geq 1$, without limit |
| Mean Euclidean Nearest Neighbor Distance Distribution (MNN) | MNN equals the mean value of distance (m) to the nearest neighboring patch of the same type | $\text{MNN} \geq 0$, without limit |
| Aggregation Index (AI) | AI equals the number of like adjacencies involving the corresponding class, divided by the maximum possible number of like adjacencies involving the corresponding | $0 < \text{AI} \leq 100$ |
| Largest Patch Index (LPI) | LPI equals the area (m2) of the largest patch of the corresponding patch type divided by total landscape area m2), multiplied by 100 | $0 < \text{LPI} \leq 100$ |
| Landscape Shape Index (LSI) | LSI equals 0.25 (adjustment for raster format) times the sum of the entire landscape divided by the square root of the total landscape area (m2) | $\text{LSI} \geq 1$, without limit |

and it is often challenging to spectrally differentiate sand from urban features. To the extent possible, we conducted field assessments to confirm the accuracy of the selected points. Further, the low level of accuracy between urban and bare soil classifications is because the materials used for building construction, including rooftops and pavement, were similar to the exposed rocks and sand in the surrounding desert areas.

## 3.4   Land-Cover Change Analysis

The results of the land-use classification process were used further to study the changes in the given period of time. To assess changes in land cover, we employed a post-classification change detection technique that required creating categorical values for each land cover. For our analysis, we assigned the following values: 1 = Urban, 2 = Vegetation, 3 = Bare soil, and 4 = Water. Subsequently, to assess the combination of possible changes (e.g., from bare soil, or 3, to urban, or (1)), a comparative analysis via a mathematical combination was developed based on five intervals: 1987–1991, 1991–1998, 1998–2003, 2003–2009, and 2009–2013, which attempted to divide the time period into roughly equal temporal segments.

The growth rate of urban land cover was quantified using the compound annual growth rate (CAGR) formula (Nassar et al. 2014):

$$CAGR = \left(\left(\frac{A(t_n)}{A(t_0)}\right)^{\frac{1}{t_n - t_0}} - 1\right) \times 100$$

where $A(t_0)$ is the initial area of urban land cover, $A(t_n)$ is the area at the end of the analysis period, and $t_n - t_0$ is the number of years covered by the analysis period. This approach was used to characterize the rate of urban growth over the particular time periods and the entire time period.

## 3.5   Urban Development Metrics

To assess the extent to which the study region is expanding, filling in, or generally changing in urban development patterns, we employed a technique of overlaying grids onto the images and quantified the land-cover patterns within each grid. While various grid sizes have been used for urban landscape analysis—150 m × 150 m (Alberti and Marzluff, 2004), 300 m × 300 m (Hart and Sailor, 2009), 500 m × 500 m (Park 1986; Unger et al. 2001), and 1 km × 1 km (Abdullah and Nakagoshi 2006)—our study area was divided into 600 m grid cells to allow each grid cell sufficient resolution for quantifying the land-cover characteristics.

Within each grid cell, we conducted an analysis using spatial landscape metrics (Gustafson 1998), which provided a quantification of the composition and configuration of the study landscapes (O'Neill et al. 2010). Using the FRAGSTATS software program (Version 4.2) (McGarigal et al. 2012), we conducted a sensitivity analysis of 13 potential landscape analysis metrics to find that 6 had the least multicollinearity ($p < 0.30$): PLAND, NP, MNN, AI, LPI, and LSI (Table 3.5). To assess temporal changes of the landscape metrics in the study region, we computed the mean value of each of the landscape metrics within each grid cell.

In addition to applying landscape metrics to each grid cell, we computed landscape metrics using the entire study site. Nassar et al. (2014) used a similar process, achieved using a time series of Landsat imagery to capture the key phases of development from 1972 to 2011, in order to quantify the urbanization process in Dubai. As Nassar employed 3 × 3 kernel size smoothing filter prior to landscape metric computation, we also applied the same filter for computing metrics for the entire site.

**Table 3.6** New categories based on mean value of PLAND

| Categories | Mean value of PLAND in each year |
|---|---|
| C1: No change | Unchanged in all years |
| C2: Early adjacent development | Rapid increase between 1987 and 1991 |
| C3: Continuous growth | Gradual increase (not year specific) |
| C4: Expansion and Infill | Rapid increase after 1998 |
| C5: Infill development | Rapid increase after 2003 |
| C6: Expansion/Urban sprawl | Rapid increase after 2009 |

## 3.6 Cluster Analysis

To develop some specific categories of urban morphology, we integrated the results of the land-cover change analysis with a statistical method called a k-means cluster analysis. The k-means algorithm is a statistical clustering method that iteratively evaluates differences among the time of development (Bacao et al. 2005). In this study, the percentage of urban land (PLAND) was computed for each 600 m grid cell, with each grid cell containing a quantifiable amount of developed (or urban) land for each study year. We then created five images that show the differences between two consecutive years for each image: 1991/1987, 1998/1991, 2003/1998, 2009/2003, and 2013/2009. These five images were the source for the k-means cluster analysis, and in this study, ten clusters were used so that the study area was divided into ten groups. We subsequently calculated the mean urban PLAND value of each cluster area for each year and plotted graphs. Those groups that had similar trends in their amount of PLAND per year were combined to create a single spatiotemporal series of six typologies of urban development in Doha: (C1) no change, (C2) early adjacent development, (C3) continuous growth, (C4) expansion and infill, (C5) infill development, and (C6) expansion/urban sprawl (Table 3.6). Figure 3.1 shows the mean urban PLAND within 600 m grid cells in each category. Each category has distinct features in terms of the trend of mean urban PLAN and spatial locations.

## 3.7 Transformation of Land Cover

Our land-cover change analysis revealed that urban development in Doha increased by approximately 315 $km^2$—or four times—during the study period (Table 3.7). The amount of vegetated area, on the other hand, has not increased by the same proportion. The data for the population of Doha was available in 1986, 1997, 2004, 2006, 2007, 2008, and 2010 (MDPS: Ministry of Development Planning Statistics 2014), and therefore we estimated the populations of other years based on the population in Qatar (World Bank 2014) by assuming the ratio between Doha and Qatar populations are similar to the closest years. The spatial extent of Doha has also changed considerably

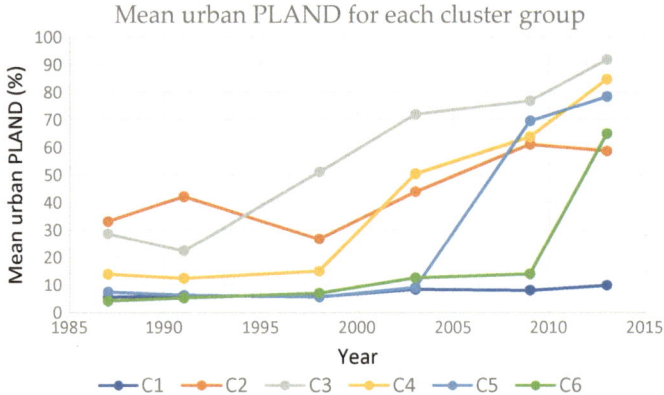

**Fig. 3.1**  Mean urban PLAND for each cluster group

**Table 3.7**  Total urban and vegetation area, population, and urban area per capita in Doha

|                                          | 1987   | 1991   | 1998   | 2003   | 2009   | 2013   |
|------------------------------------------|--------|--------|--------|--------|--------|--------|
| Total urban (km$^2$)                     | 106.02 | 113.24 | 120.23 | 225.80 | 284.76 | 421.17 |
| Total vegetation (km$^2$)                | 11.19  | 7.47   | 18.76  | 25.58  | 13.84  | 22.01  |
| Ratio between vegetation and urban (%)   | 10.56  | 6.60   | 15.61  | 11.33  | 4.86   | 5.23   |
| Estimated population (in thousand)       | 225    | 246    | 274    | 322    | 760    | 991    |
| Urban area per capita (m$^2$)            | 470    | 460    | 439    | 702    | 375    | 425    |

over the period of study (Fig. 3.2). Specifically, the amount of urban area has rapidly increased, especially along Salwa Highway to the Southwest and Doha Expressway to the Northwest. In addition, a considerable alteration in the form of the coastline occurred at two distinct places, The Pearl and Hamad International Airport, between 2003 and 2009.

## 3.8  Urban Growth Dynamics

Changes in the composition and configuration of urbanizing Doha were found by examining the mean value of the landscape metrics within a grid cell (Fig. 3.3). Likely due to the consolidation of urban development, the PLAND metric steadily increased during the study period, while the number of patches (NP) decreased after 1998. The aggregation index (AI) steadily increased after 1998, which further suggests that the region is becoming less fragmented and more agglomerated with urban development.

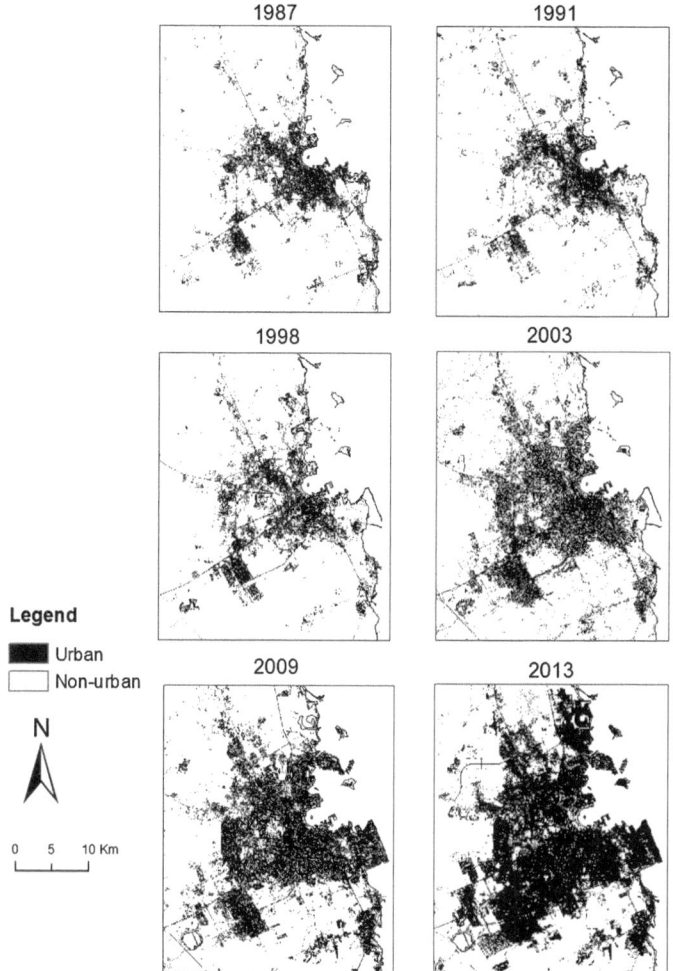

**Fig. 3.2**  Change in urban extent in Doha in 1987, 1991, 1998, 2003, 2009, and 2013

## 3.9   Clustering Land Covers

The land-cover change analysis provided a means of spatially articulating six pattern categories of urban growth. While the core areas of the West Bay, the central business district (CBD), and the outer edges remained unchanged over the study period, all of the other areas could be placed into one of five development patterns (Fig. 3.4). Early adjacent development (C2, 73 km$^2$) was sporadically distributed in areas that were developing (or being developed) in 1987, the first year of our study period. A 48 km$^2$ continuous growth development (C3) occurred over the study period, mostly located between the CBD and expansion/urban sprawl development. Expansion and

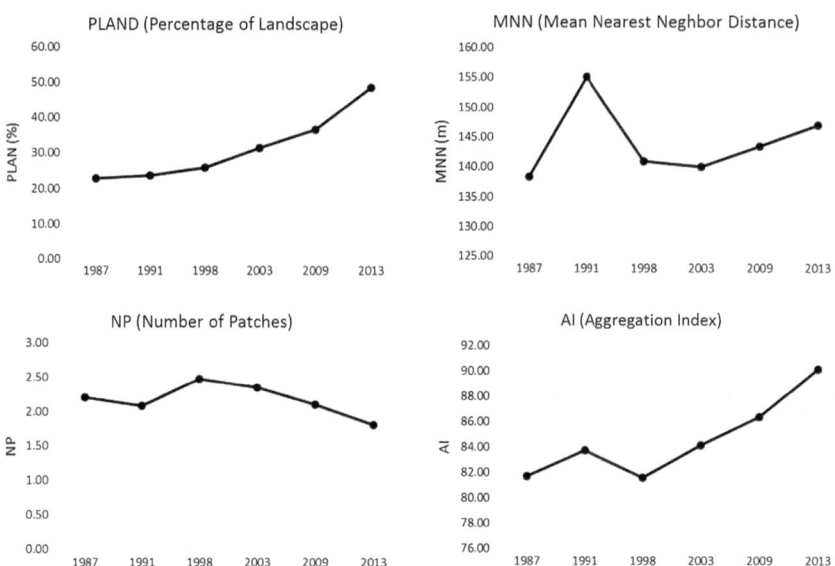

**Fig. 3.3**  Temporal change of select landscape metrics for Doha

infill development (C4, 145 km$^2$) located mostly adjacent to continuous growth development, but its rapid growth took place after 1998. Infill development (C5, 38 km$^2$) occurred after 2003, and located around the Pearl and the New Hammad airport. Finally, the largest growth of urban development pattern over the study period was the expansion (C6, 234 km$^2$) that occurred mostly on the outskirts of the city after 2009.

## 3.10  Implications for the Region

While earlier studies have focused on the historical and cultural changes in Doha since the development of the oil infrastructure (see, for example, Wiedmann et al. 2012; Wiedmann and Salama 2012), we focused on the creation of physical patterns that were introduced through new models of urbanism that focused on urban growth and megaprojects. Earlier studies confirmed that Doha's current stage of urban development patterns is a product of many co-occurring factors, including the creation of an international hub for education and sports, and liberalization and decentralization process to accelerate urban growth (Wiedmann and Salama 2012). Many of the patterns in Doha are akin to the models of urban growth and change in the U.S. and Europe. This expansion comes in the form of low-density suburban development relatively far from the original areas where Doha was founded. Other development seems to be large megaprojects that are distributed throughout the metro region, such as the Hammad Airport and the Pearl. Additionally, while much of the expanding

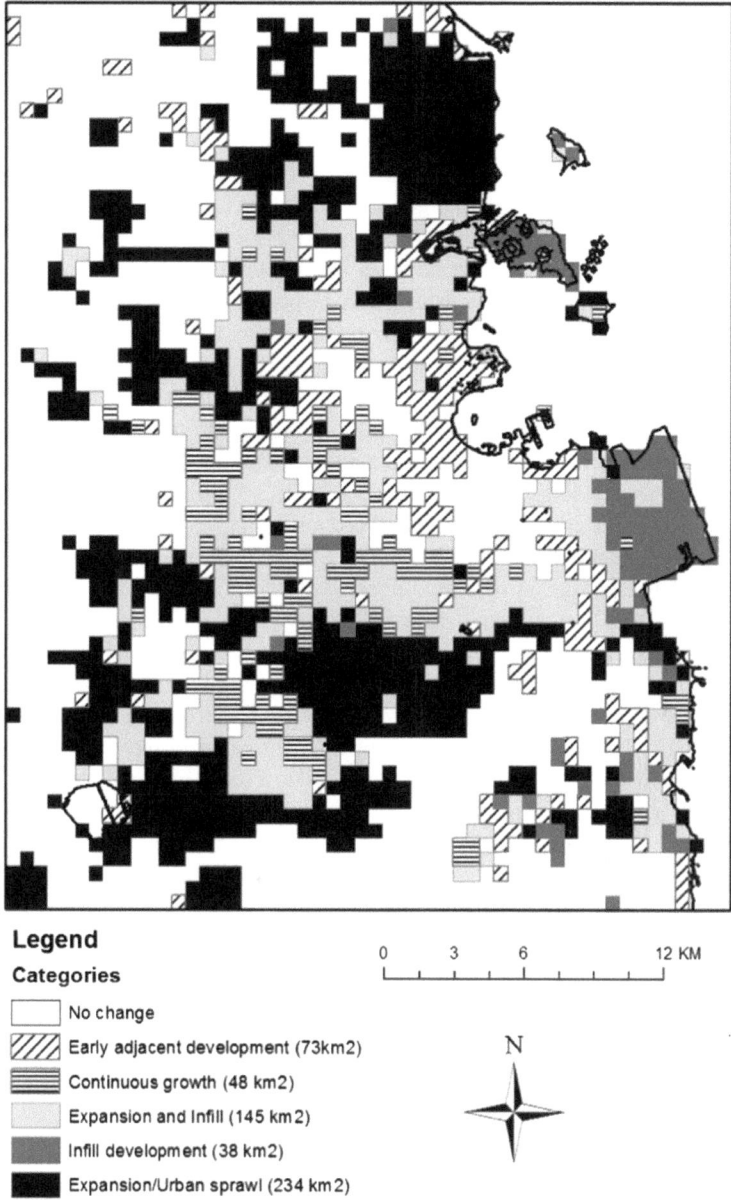

**Legend**

**Categories**

☐ No change

▨ Early adjacent development (73km2)

☰ Continuous growth (48 km2)

▨ Expansion and Infill (145 km2)

▨ Infill development (38 km2)

■ Expansion/Urban sprawl (234 km2)

0    3    6         12 KM

N

**Fig. 3.4** Results of k-means cluster analysis, combined to aforementioned six categories

development of the North American and Western European cities occurred over the course of a century, Doha has seen almost the equivalent growth in 20 years.

In terms of observed land-cover change, the patterns of growth were largely urban, which grew four times ($106–421$ km$^2$) over the study period. Our results indicate that early in the process of development, between 1987 and 1998, the Doha region grew outward, resulting in an overall increase in the distance between development areas (as measured by Mean Nearest Neighbor, MNN). The region has also infilled during the study period, as evidenced by an increase in the aggregation index (AI). Increases in AI suggest that formerly disparate areas were brought together by the development of the urban infrastructure. This pattern may be part of the new development districts with distinctive architectural characteristics, such as the Pearl, Katara, and West Bay. While suburban development in North American and Western Europe generally fragments the surrounding forests or farmlands, a different effect has occurred in growth in Doha. Since the areas surrounding the city do not contain vegetation, for example, new development has created a modest overall increase in vegetation. The total vegetation in 1998 was approximately 19 km$^2$, 1.7 times that of 1987, but only slight increase to 22 km$^2$ by 2013. The proportion of vegetation in comparison to urban area decreased from 11% in 1987 to 5% in 2013. Statistical analysis further reveals that the expansion of development mentioned earlier, which we refer to as "Early Adjacent Development," occurred mostly outside of the urban core.

In terms of a sustainable urban pattern, two characteristics can be separated out in the evolution of the Doha since 1987. First, the initial development was outward and relatively low density, which follows many patterns from the U.S. and European cities. From 1987 through 2003, the development came in the form of expansion across away from the central West Bay area, and to a lesser extent along the coastline. These development patterns encouraged automobile use, and without limited public transit creates an increase in air pollution, infrastructure costs, and obesity. Indeed, recent studies are beginning to shed light on the challenges of these land-use patterns, including the world's worst air quality (WHO 2016), and the most obesity within the Middle East and North Africa (Ng et al. 2014). Some argue that megaprojects, including Education City, Al-Waab, the Aspire Zone, located far from the city center helped to spurn the rapid expansion and sprawling land-use patterns (Rizzo 2013). Since our analysis did not examine megaprojects, we are unable to confirm these conclusions; however, we note that early developments occurring from the coastline established a periphery of the metropolitan region, which served as a boundary that may have in later years allowed for infill development.

A second set of land-use patterns suggest many indications for creating a more sustainable Doha. The infill development that has been occurring since 2003, including the high densities of the Pearl and Msheireb in downtown, contains many core components of sustainable land use. In fact, Qatar's national framework for urban development describes the need for greater connectivity among land uses, greater densities, including corridors and centers, and opportunities for infill development. The recent creation of the transit line in Doha is expected to cover over 200 km is a compelling example of the land-use strategy taking root. The metro rail lines are expected to have their hub at Msheireb, with a total of four lines connecting all the major areas from the periphery to the core. Such efforts, if coupled with opportunities for greater neighborhoods that allow for active transit and walkability, more vegetation, and outdoor spaces, can advance livability as mounting pressures from climate change further threaten unstainable land-use patterns. Overall, while the land-cover patterns of Doha reveal dynamic and rapid change over the study period, the urban growth patterns support previous descriptions of expansion and infill, although with specific temporal patterns. What has taken many cities in the U.S. over 70 years of expansion and infill has taken Doha only 25. Indeed, our analysis reveals that while the initial growth of Doha spurred unstainable land-use patterns, more recently, the City has created many opportunities for sustainable growth.

While the physical growth of Doha is discernible through an analysis of satellite images, the quality of growth and its implications for the social, economic, and environmental conditions of the region all require further study. Our study cannot, in its current state, shed light on how the physical changes are affecting Doha residents. We can, however, speculate from other studies that the patterns of initial suburbanization have required greater investment in infrastructure and impacted the capacity for residents to engage in active transportation (e.g., biking, walking, transit, etc.). These will further have implications on public health and the environment, which we have observed in other parts of the world. Needed are examinations of Doha's development patterns that are linked to environmental factors and overall quality of life, if for no other reason than to understand whether development patterns in and of themselves create similar outcomes in other cultures. At best, such future research could help to reduce the possible negative implications of development for humans and the environment upon which they depend.

# References

Abdullah SA, Nakagoshi N (2006) Changes in landscape spatial pattern in the highly developing state of Selangor, Peninsular Malaysia. Landsc Urban Plan 77(3):263–275. https://doi.org/10.1016/j.landurbplan.2005.03.003

Alberti M (2010) Maintaining ecological integrity and sustaining ecosystem function in urban areas. Curr Opin Environ Sustain 2(3):178–184

Alberti M, Marzluff J (2004) Resilience in urban ecosystems: Linking urban patterns to human and ecological functions. Urban Ecosyst 7:241–265

Bacao F, Lobo V, Painho M (2005) Self-organizing maps as substitutes for K-means clustering. Lect Notes Comput Sci 3516:476–483

Bertaud A, Renaud B (1997) Socialist cities without land markets. J Urban Econ 41(1):137–151. https://doi.org/10.1006/juec.1996.1097

Bhatta B, Saraswati S, Bandyopadhyay D (2010) Urban sprawl measurement from remote sensing data. Appl Geogr (Clim Chang Appl Geogr Place Policy Pract) 30(4):731–740. https://doi.org/10.1016/j.apgeog.2010.02.002

Brower BV, Laben CA (2000) U.S. Patent No. 6,011,875

Brueckner JK (2001) Urban sprawl: lessons from urban economics. In: Gale WG, Pack JR (eds) Brookings-Wharton papers on urban affairs. Brookings Institution, Washington D.C., pp 65–89

Cho J (2005) Urban planning and urban sprawl in Korea. Urban Policy Res 2:203–218

Cutsinger J, Gastler G (2006) There is no sprawl syndrome: a new typology of metropolitan land use patterns. Urban Geogr 27(3):228–252

Ewing R, Cervero R (2001) Travel and the built environment: a synthesis. Transp Res Rec J Transp Res Board 1780:87–114

Fenkel A, Ashkenazi M (2008) Measuring urban sprawl; how can we deal with it? Environ Plan 35(1):56–79

Foody GM (2002) Status of land cover classification accuracy assessment. Remote Sens Environ 80(1):185–201. https://doi.org/10.1016/S0034-4257(01)00295-4

Frumkin H (2002) Urban sprawl and public health. Public Health Rep 117(3):201–217. https://doi.org/10.1093/phr/117.3.201

Glaeser EL, Kahn ME (2003) Sprawl and urban growth. Working paper series

Gustafson EJ (1998) Minireview: quantifying landscape spatial pattern: what is the state of the art? Ecosystems 1(2):143–156. https://doi.org/10.1007/s100219900011

Hart MA, Sailor DJ (2009) Quantifying the influence of land-use and surface characteristics on spatial variability in the urban heat island. Theoret Appl Climatol 95(3):397–406. https://doi.org/10.1007/s00704-008-0017-5

Hortas-Rico M, Solé-Ollé A (2010) Does urban sprawl increase the costs of providing local public services? Evidence from Spanish municipalities. Urban Stud 47(7):1513–1540. https://doi.org/10.1177/0042098009353620

Li F, Liu X, Hu D, Wang R, Yang W, Li D, Zhao D (2009) Measurement indicators and an evaluation approach for assessing urban sustainable development: a case study for China's Jining City. Landsc Urban Plan 90(3):134–142. https://doi.org/10.1016/j.landurbplan.2008.10.022

Li F, Peijun D (2013) Dynamic change analysis of urban sprawl based on remote sensing and GIS-a case study of Jiangning, Nanjing, China. In: Proceedings of the JURSE. Wiley, New York

Lillesand TM (2004) Remote sensing and image interpretation, 5th edn. Wiley, New York

Lo CP, Choi J (2004) A hybrid approach to urban land use/cover mapping using Landsat 7 Enhanced Thematic Mapper Plus (ETM+) images. Int J Remote Sens 25(14):2687–2700

Loutzenheiser D (1997) Pedestrian access to transit: model of walk trips and their design and urban form determinants around bay area rapid transit stations. Transp Res Rec J Transp Res Board 1604:40–49. https://doi.org/10.3141/1604-06

McGarigal K, Cushman S, Ene E (2012) FRAGSTATS v4: spatial pattern analysis program for categorical and continuous maps. Computer software program produced by the authors at the University of Massachusetts, Amherst

MDPS: Ministry of Development Planning Statistics (2014) Statistics tree

Monkkonen P (2008) Using online satellite imagery as a research tool: mapping changing patterns of urbanization in Mexico. J Plan Educ Res 28(2):225–236. https://doi.org/10.1177/0739456X08323771

Nassar AK, Alan Blackburn G, Duncan Whyatt J (2014) Developing the desert: the pace and process of urban growth in Dubai. Comput Environ Urban Syst 45:50–62. https://doi.org/10.1016/j.compenvurbsys.2014.02.005

Ng M, Fleming T, Robinson M, Thomson B, Graetz N, Margono C, Abraham JP et al (2014) Global, regional, and national prevalence of overweight and obesity in children and adults during 1980–2013: a systematic analysis for the Global Burden of Disease Study 2013. The Lancet 384(9945):766–781

O'Neill BC, Dalton M, Fuchs R, Jiang L, Pachauri S, Zigova K (2010) Global demographic trends and future carbon emissions. Proc Natl Acad Sci 107(41):17521–17526. https://doi.org/10.1073/pnas.1004581107

Park H-S (1986) Features of the heat island in seoul and its surrounding cities. Atmos Environ (1967) 20(10):1859–1866. https://doi.org/10.1016/0004-6981(86)90326-4

Parsons Brinckerhoff Quade, Douglas Inc. (1994) Building orientation: a supplement to the pedestrian environment

Radhakrishnan N (2014) Analysis of urban sprawl pattern in Tiruchirappalli city using applications of remote sensing and GIS. Arab J Sci Eng 7:5555–5563

Report U (2012) The state of Arab cities 2012: challenges of urban transition. UN Habitat, Nairobi, Kenya

Rizzo A (2013) Metro Doha. Cities 31:533–543. https://doi.org/10.1016/j.cities.2011.11.011

Schowengerdt RA (1997) Remote sensing: models and methods for image processing. Elsevier

Shalaby A, Ali R, Gad A (2012) Urban sprawl impact assessment on the agricultural land in Egypt using remote sensing and GIS: a case study Qalubiya Governorate. J Land Use Sci 7(3):261–273

Song Y, Knaap G-J (2004) Measuring urban form: is portland winning the war on sprawl? J Am Plan Assoc 70(2):210

Unger J, Sümeghy Z, Gulyás Á, Bottyán Z, Mucsi L (2001) Land-use and meteorological aspects of the urban heat island. Meteorol Appl 8(2):189–194. https://doi.org/10.1017/s1350482701002067

WHO (World Health Organization) (2016) Global ambient air pollution database. http://www.who.int/phe/health_topics/outdoorair/databases/cities/en/. Accessed 10 July 2016

Wiedmann F, Salama AM (2012) Urban evolution of the city of Doha: an investigation into the impact of economic transformations on urban structures. METU JFA 29(2):35–61

Wiedmann F, Salama AM, Thierstein A (2012) Urban evolution of the city of Doha: an investigation into the impact of economic transformations on urban structures. Metu J Fac Arch 29(2):35–61. https://doi.org/10.4305/METU.JFA.2012.2.2

Wilby RL (2007) A review of climate change: impacts on the built environment. Built Environ 33(1):31–45. https://doi.org/10.1111/j.1523-1739.2009.01264.x

World Bank (2014) Data: Qatar. Resource document. http://data.worldbank.org/country/qatar. Accessed 22 Aug 2018

# Chapter 4
# Regional Variations in Temperatures

Vivek Shandas, Yasuyo Makido and Salim Ferwati

**Abstract** Recent evidence suggests that urban forms and materials can help to mediate temporal variation of microclimates and those landscape modifications can potentially reduce temperatures and increase accessibility to outdoor environments. To understand the relationship between urban form and temperature moderation, we examined the spatial and temporal variation of air temperature throughout one desert city—Doha, Qatar—by conducting vehicle traverses using highly resolved temperature and GPS data logs to determine spatial differences in summertime air temperatures. To help explain near-surface air temperatures using land-cover variables, we employed three statistical approaches: ordinary least squares (OLS), regression tree analysis (RTA), and random forest (RF). We validated the predictions of the statistical models by computing the root mean square error (RMSE) and discovered that temporal variations in urban heat are mediated by different factors throughout the day. The average RMSE for OLS, RTA, and RF are 1.25, 0.96, and 0.65 (in Celsius), respectively, suggesting that the RF is the best model for predicting near-surface air temperatures at this study site. We conclude by recommending the features of the landscape that have the greatest potential for reducing extreme heat in arid climates.

**Keywords** Urban Heat Island · Doha · Statistical methods · Models

While several emerging studies of desert cities have identified those landscape features that mediate urban heat, we currently lack the ability to describe differences in temperatures throughout the day, in part because earlier studies relied on remotely sensed descriptions of cities, which lack the ability to describe the diurnal profiles of temperature. Early studies using satellite imagery began decades ago with Rao (1972), who examined surface urban heat islands from thermal infrared data. Since

---

Sections of this chapter are from the following document, which is part of the Creative Commons Open Licensing system: Makido, Y, V Shandas, S Ferwati, and D Sailor, 2016. Daytime Variation of Urban Heat Islands: The Case Study of Doha, Qatar. *Climate (4)*, 32.

then, various sensor platforms—such as NOAA, AVHRR, Landsat MSS, TM, and ASTER—have been used to determine the urban thermal climate. More recently, land surface temperatures (LST) derived from thermal bands of radiometric instruments flown on satellites and high-altitude aircraft have been used to analyze the relationships among LST and landscape characteristics such as NDVI, NDBI, percent impervious surface, and percent vegetation by using linear regression (Yuan and Bauer 2007; Li et al. 2011), multiple linear regression (Frey et al. 2007), geographically weighted regression (Su et al. 2012), and correlation analysis (Zhou et al. 2014).

While such studies are helpful in guiding initial heat mitigation efforts, they are quite general and face several challenges which make them not wholly useful in practice. For example, areas vary in the composition and configuration of landscape features, and recommending greening in an industrial area that contains little planting spaces may prove a challenge due to the lack of available planting space. Similarly, studies of heat mitigation in the built environment often refer to urban areas as homogeneous units, with limited differentiation between intra-urban landcover types (Santamouris et al. 2017), making application of mitigation techniques difficult or ineffective at best. To guide the practice of urban heat mitigation efforts, practitioners need models that (1) are adaptable to distinct cities and land uses, providing thresholds of mitigation potential; and (2) describe land-cover types within the city and thus indicate a variety of best intervention strategies, rather than one generalized approach.

One approach to developing a temporal-based assessment of urban heat is to use ground-based empirical measures of temperature. One advantage of using ground-based measurements is the capacity to describe differences that individuals and communities directly experience. Oke (1976) measured air temperature from car traverses in Vancouver, B.C. and claimed that heat island intensity is more directly related to the physical structure of an urban area than its size. More recently, Hart and Sailor (2009) employed tree-structured regression models to investigate the spatial variability of urban heat intensity using vehicle temperature traverse and GIS resources in Portland (Oregon, USA). Tree-structured regression models enable the determination of the most important land-use and surface variables affecting the UHI intensity of metropolitan areas. Later, Heusinkveld et al. (2014) conducted bicycle traverse measurements to assess the spatial variation of temperature during a summer day in Rotterdam. They found that spatial variations in temperature were strongly affected by local vegetation cover. Using a ground-based approach and regression analysis, Yan et al. (2014) found that localized land-cover composition and site geometry were two of the most important variables affecting local air temperatures in Beijing. Further, Ho et al. (2014) examined three statistical models by calibrating satellite-derived predictors such as LST and weather station data to map the daily maximum air temperature distribution in greater Vancouver. They found that the random forest model was the best among three statistical models used to map temperature distributions in the area. Finally, Yokobori and Ohta (2016) used mobile traverses to conduct air temperature observations that would clarify the effect of land cover on ambient air temperatures in Tokyo. They found that an intra-urban heat island existed throughout the year, and they observed that air temperatures varied significantly according to ambient land-cover types.

The use of ground-based and remote measurements of urban temperatures is a promising step toward better describing the spatial variation; however, several questions remain unaddressed. For example, satellite measurements can offer periodic descriptions of a city, yet temperatures vary throughout the day, and "snapshots" of surface temperatures preclude an understanding of how those factors affect temporal variations in urban heat. While the land-cover variables derived from the satellite images do not change throughout the day, we hypothesized that the effects of these variables will change throughout the day as temperatures change. In addition, it is unknown what techniques might allow for the higher predictability of urban heat in cities that have limited land-use and land-cover information available. To better understand the relationship between temperatures, diurnal cycles, and built environment factors, we examined the spatial and temporal variation of air temperatures throughout one city—Doha, Qatar. We addressed two research questions in this study: (1) which statistical models best explain the variability of urban heat? and (2) what land-use and land-cover factors best describe the variations in urban heat island effects throughout the day? To address these questions, we used a combination of ground-based, empirically derived, mobile vehicle temperature traverses and land-cover features that helped to populate three types of statistical models: ordinary least squares, regression trees, and random forests.

## 4.1  Measuring Near-Surface Air Temperatures

We used a cross-sectional research design to capture variations of near-surface air temperatures (referred to as "air temperature" in this paper) throughout the day in Doha, Qatar. Numerous vehicle traverses were the primary means used to assess both spatial and temporal temperature variations in the study area. Based on these empirically derived data, we developed statistical models to evaluate the extent to which specific land-use variables helped to describe variations in temperature. We begin here by describing the vehicle traverses and land-cover variables, which provide the context for subsequent descriptions of our statistical models.

## 4.2  Vehicle Temperature Traverses

The vehicle traverses examined in this study on September 8 and 9, 2014 and May 12, 14, and 15, 2015 covered all parts of the city of Doha (Fig. 4.1). Following an established protocol (Hart and Sailor 2009), we used a Type T fine (30 gauge) thermocouple mounted in a 12-cm-long, 2.5-cm-diameter white plastic shade tube. The tube was supported approximately 25 cm above the vehicle roof on the passenger side window (Fig. 4.2). The temperature sensors were connected to data logging

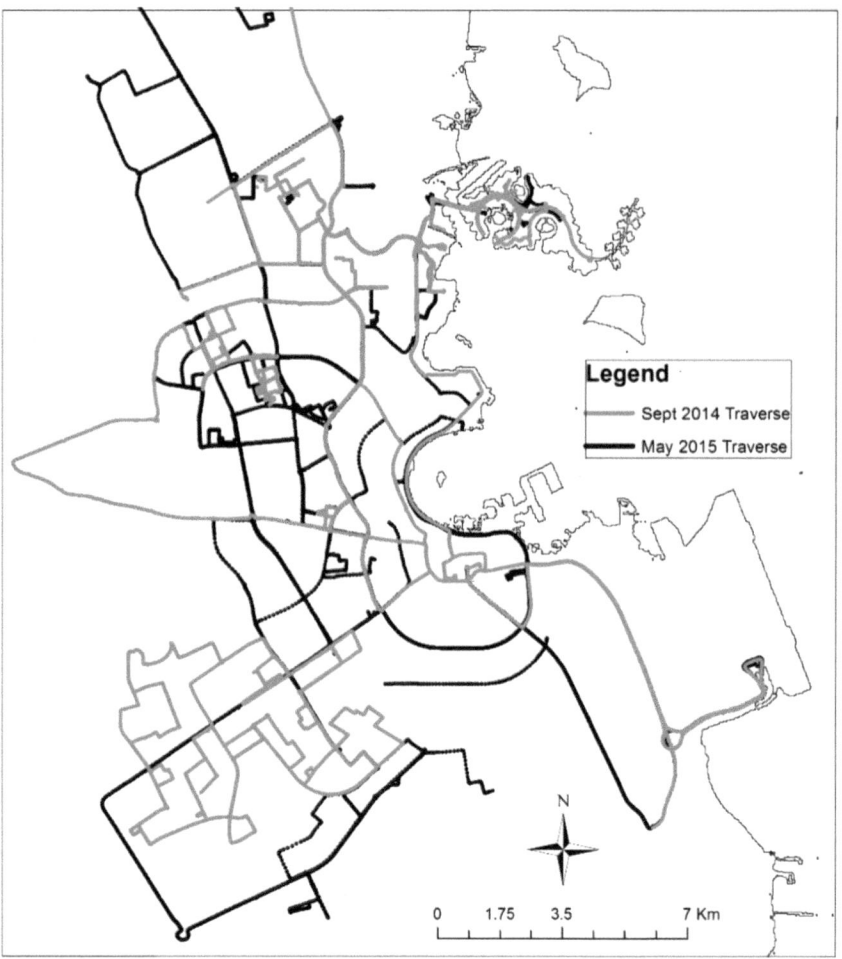

**Fig. 4.1** Vehicle temperature traverses across the study area, Doha. Coverage on September 8 and 9, 2014 and May 12, 14, and 15, 2015

temperature recorders with an estimated system accuracy of ±0.5 °C and a 90% response time of less than 60 s at 1 m/s airflow. A time-synchronous GPS system was also attached to each car so that the temperature measurements (with a sampling frequency of 1 s in September 2014 and 10 s in May 2015) could be paired with a GPS location. Data for vehicle speeds of less than 5 km/h were discarded as the temperature sensors were aspirated by the movement of the vehicle, and we aimed to avoid oversampling when the vehicles were stopped (e.g., in traffic or at traffic lights). Each day's traverse involved four cars, lasted one hour, and was conducted at three time periods: 6:00–7:00, 13:00–14:00, and 19:00–20:00. Each car traversed approximately 25–30 km.

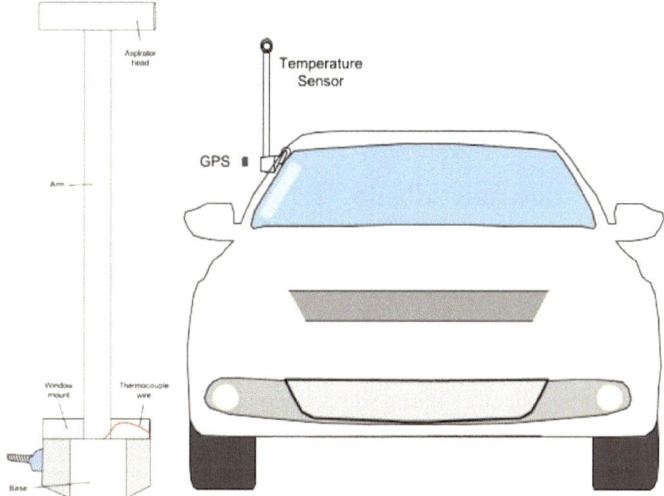

**Fig. 4.2** Temperature sensor mounted within plastic tube (left) and vehicle with the plastic tube (right). Image by Jackson Voelkel

## 4.3  Land-Use and Land-Cover Variables

We acquired 30-m-resolution satellite sensor imagery from the US Geological Survey's Landsat OLI files taken on September 12, 2014. The study area (23.5 km × 25.9 km), which encompassed the Doha metropolitan region, was extracted from the scene of the satellite sensor imagery. All Landsat bands (visible, near, and short-wave infrared) except the thermal band were used in the classification process. For accuracy assessment, due to a lack of ground truth data at this study site, we used existing QuickBird™ high-resolution imagery (acquisition date = April 3, 2013). Since the acquisition dates of Landsat and QuickBird were separated by nearly one and a half years, we examined how the landscapes changed over time. We employed change detection analysis using two images: Landsat 2013 (acquisition date = July 3, 2013) and Landsat 2014 (acquisition date = December 9, 2014). For this study, we examined two types of feature indices: vegetation index (NDVI, Eq. (4.1)) (Jensen 1986) and built-up index (NDBI, Eq. (4.2)) (Zha et al. 2003).

$$NDVI = (Band\ 5 - Band\ 4)/(Band\ 5 + Band\ 4) \tag{4.1}$$

$$NDBI = (Band\ 6 - Band\ 5)/(Band\ 6 + Band\ 5) \tag{4.2}$$

The result of the change of detection analysis suggested that the change of vegetation and built-up areas occurred in less than 0.3% of the study area, which validated the QuickBird™ image as acceptable ground truth data. All image processing work and accuracy assessments were carried out in ENVI 5.1 (Exelis Visual Information Solutions, Boulder, CO, USA) and ArcGIS 10.2 (ESRI Inc., Redlands, CA, USA).

## 4.4  Classification and Albedo

In order to compensate for the lack of ancillary data at this study site, and based on the prior satellite-based research of the study area (Shandas et al. 2016), we employed a simplified hybrid classification method (Lo and Choi 2004). Each image, with six bands, was sorted into 40 clusters using the ISODATA unsupervised classification method. Only homogeneous clusters were labeled into one of four categories: Urban, vegetation, soil, and water (Table 4.1). Other clusters that exhibited mixed classes were clipped out from the image for further classification using supervised classification. The maximum likelihood algorithm, which evaluates both the variance and covariance of the category spectral response patterns when classifying an unknown pixel, was applied to the clipped images (Lillesand et al. 2014). The images were initially classified into seven classes: Urban 1, Urban 2, Vegetation, Soil 1, Soil 2, Water 1, and Water 2. Ten to 20 training sites for each class were selected with the aid of panchromatic band and/or QuickBird™ imagery. The classified images were resampled to the aforementioned four classes and combined with the ISODATA classified land-use/land-cover map.

Albedo was computed from visible and near-infrared bands in Landsat 8 using Eq. (4.3) (Yale Center for Earth Observation, no date). For this equation, Liang's [30] formula to calculate Landsat shortwave albedo was normalized (Smith 2010). Although this formula was developed for Landsat ETM+, it is applicable to Landsat OLI [32] as the band numbers were adjusted to conform to the OLI standard

$$\alpha_{short} = \frac{0.356\rho_2 + 0.130\rho_4 + 0.373\rho_5 + 0.085\rho_6 + 0.072\rho_7 - 0.0018}{0.356 + 0.130 + 0.373 + 0.085 + 0.072} \quad (4.3)$$

where $\rho$ represents the top of atmosphere (TOA) reflectance of Landsat bands 2, 4, 5, 6, and 7. The resulting digital number was converted to TOA using the radiometric calibration tool in ENVI 5.1.

**Table 4.1**  Land-use/land-cover classes and definitions used in this study

| Classes | Definition |
| --- | --- |
| Urban | All built-up surfaces, including roads, commercial, industrial pavements, and construction sites |
| Vegetation | All areas of vegetation, including farms, parks golf courses, and lawns |
| Soil | Bare and exposed rock, coastal sands, and sand dunes |
| Water | All areas of open water, including lakes and the ocean |

## 4.5  Accuracy Assessment

To address the lack of ground-level land-use/land-cover data, we conducted an accuracy assessment using high-resolution satellite sensor imagery with half of the total images that we classified. To conduct an accuracy assessment of the Landsat images, the QuickBird™ imagery was ortho-rectified and pan-sharpened. Orthorectification is the correction of the image, pixel-by-pixel, for topographic distortion and results in every pixel appearing to be viewed from directly above (Schowengerdt 1997). Pan-sharpening, meanwhile, is a process of combining a lower resolution multispectral imagery with high-resolution panchromatic imagery to create a high-resolution color image. For this study, Gram-Schmidt pan-sharpening (Brower and Laben 2000) was applied to obtain 60-cm multispectral data. All images were georeferenced using first-order polynomial transformation, and the overall root mean square errors were less than half a pixel.

We also used a standard confusion matrix method by randomly sampling approximately 240 points (about 60 points for each of the four classes) and laying those points over the QuickBird™ imagery. We categorized each point into one of the four classes by visually determining the dominant land-cover types within 30 × 30 m polygons that coincided with the Landsat imagery's pixel footprints. The categorized data were used as reference data and the matrix was created to compare the relationship between the reference data and the corresponding classified data (Table 4.2). The overall classification accuracy was 85.6% and the kappa coefficient was 0.81, which is consistent with the standard minimum accuracy of 85 (Anderson et al. 1976). Some of the 15% misclassifications consist of spectral differentiation between sand/soil and other urban features (e.g., rooftops and parking lots that consist of sand and/or soil).

**Table 4.2**  Confusion matrix of accuracy assessment of the land-cover classification for 2014

|  |  | Reference data | | | | | User's accuracy (%) |
|---|---|---|---|---|---|---|---|
|  |  | Urban | Vegetation | Soil | Water | Total |  |
| Classified data | Urban | 63 | 1 | 4 | 0 | 68 | 92.7 |
|  | Vegetation | 8 | 47 | 4 | 0 | 59 | 79.7 |
|  | Soil | 7 | 1 | 52 | 0 | 60 | 86.7 |
|  | Water | 7 | 1 | 1 | 40 | 49 | 81.6 |
|  | Total | 85 | 50 | 61 | 40 | 236 |  |
| Producer's accuracy (%) |  | 74.1 | 94.0 | 85.3 | 100.0 |  |  |

## 4.6  Predicting Air Temperatures

Each 30-m pixel was further evaluated in terms of four additional land surface variables: urban/nonurban, vegetation/non-vegetation, distance to the coast, and albedo. Urban, vegetation, and water classes, which were derived through hybrid classification, were used for these variables. Based on Heusinkveld et al.'s (2014) approach of using a linear regression correlation coefficient between urban temperature and greenery density as a function of pie-shaped distance (ArcMap 10.x, geostatistical analyst), we quantified the area of wind-induced influence on each of the land surface variables for each discrete temperature measurement. We did not find a directional effect on the correlation coefficient, so we employed a circular-shaped area (called a buffer) and tested buffer distances between 50 and 600 m at 50 m increments. Based on the correlations found between buffer size and temperature, we selected the most influential buffer sizes for each date/time by determining the buffer distances with the largest positive or negative effects on temperatures (Table 4.3). The maximum buffer size of 600 m was selected based on previous studies, which commonly employed several hundred meters (Krüger and Givoni 2007) and 500 m/1000 m (Ho et al. 2016) as buffer distances. The land surface variables employed, and the strongest distance effects, were as follows:

- Mean albedo within a certain radius (a#),
- Percentage of urban area within a certain radius (u#),
- Percentage of vegetation cover within a certain radius (v#), and
- Distance to the coast (w_dist).

We employed three statistical approaches to model air temperatures using these relevant predictors: ordinary least squares (OLS), regression tree analysis (RTA), and random forest (RF). OLS is the most commonly employed approach to understand the strength of independent variables in order to explain dependent variables (Frey et al. 2007; Yuan and Bauer 2007; Li et al. 2011).

RTA offers an alternative technique for handling nonlinear relationships between the dependent variable and the predictive variables (Yuan et al. 2008). The RTA algorithm uses a set of independent variables to recursively split dependent variables into subsets, which maximizes the reduction in the residual sum of squares (Hansen et al. 2002). Based on conditional probabilities, the tree contains left and right nodes. In our case, the left node indicated that a condition was true while the right node indicated that it was false (Fig. 4.3). The study area was divided into four to six categories based on the node criteria. The values at each terminating node were considered for the mean temperature for that terminal node and served as an input for the multiple linear regression model. Depending on the specific parameters of each pixel in the study area, the regression tree was constructed to predict the temperature of each pixel. The first terminating node indicated the highest explanatory power for determining urban heat. While the specific regression trees varied by time of day— for May 12 at 1 pm, for example, most of the variation of temperature was based on the distance to coast—the regression tree analysis offered a visual and analytical approach to describing factors that affect local variations in temperature.

**Table 4.3** Buffer sizes for each variable (in meters)

| | September 8, 2014 | | | September 9, 2014 | | |
|---|---|---|---|---|---|---|
| | 6 am | 1 pm | 7 pm | 6 am | 1 pm | 7 pm |
| Albedo | 50 | 400 | 150 | 50 | 400 | 50 |
| Urban | 500 | 600 | 600 | 600 | 600 | 600 |
| Vegetation | 100 | 400 | 150 | 50 | 350 | 150 |

| | May 12, 2015 | | | May 14, 2015 | | | May 15, 2015 | | |
|---|---|---|---|---|---|---|---|---|---|
| | 6 am | 1 pm | 7 pm | 6 am | 1 pm | 7 pm | 6 am | 1 pm | 7 pm |
| Albedo | 150 | 600 | 600 | 50 | 400 | 600 | 50 | 600 | 200 |
| Urban | 600 | 50 | 50 | 100 | 600 | 300 | 50 | 600 | 600 |
| Vegetation | 50 | 300 | 600 | 600 | 100 | 200 | 50 | 600 | 600 |

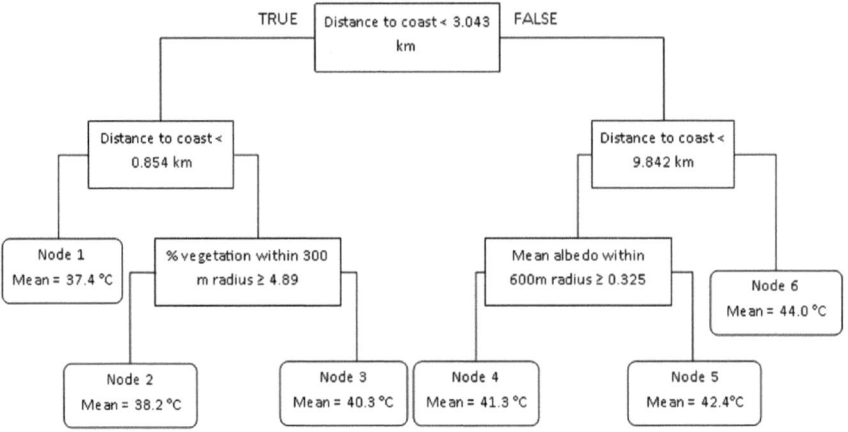

**Fig. 4.3** Example of regression tree analysis with terminal nodes for May 12, 1 pm

RF, a machine learning technique, is one of the data mining methods designed to produce accurate predictions that do not overfit the data. In a random forest analysis, a group of regression trees is created using samples from the training data, which is called "bagging." Unlike a regression tree, a random forest tree splits each node by using the best of a randomly chosen subset at that node. For the regression, the average prediction for all trees are considered; for classification, predictions considered the "vote" for the most popular class among all trees (Breiman 2001; Liaw and Wiener 2002; Prasad et al. 2006). RF has been used with satellite-derived predictors to map air temperature distributions (Ho et al. 2014, 2016). We used RF to test variables for both selected buffer sizes and all buffer sizes. Based on preliminary studies, the number of variables randomly sampled as candidates at each split was set to 20 for all variables and, for selected variables, to four. We selected a total of 500 trees to be grown. We used the R statistical software (R Development Core Team, no date) for OLS, an additional extension called "rpart" for the regression tree analysis (Therneau et al, no date) and "randomForest" for the random forest analysis (Liaw and Wiener 2002).

In order to compare the accuracy of each statistical model, we employed a "holdout method," which partitions the data into two mutually exclusive subsets called a training set and a test (i.e., holdout, set). Common applications of the holdout method suggest selecting 2/3 of the data as the training set and the remaining 1/3 as the test set (Kohavi 1995). For this study, we used 70% of the traverse data, randomly selected, as the training set and the remaining 30% of the data as the test set.

## 4.7  Distribution of Air Temperatures

Using the results of the three models, we developed spatial descriptions of temperature variability for each date/time. The prediction of the air temperature map, using the three statistical models on May 12, 2015, 1 pm vehicle traverse data, showed considerable variation in the results (Fig. 4.4). Specifically, while all three models indicated a gradient of cooler to hotter moving inland, the OLS model seemed to generalize the differences, while the Random Forest (RF) model indicated sharp contrast within smaller units across the study region. The regression tree analysis indicated a notable and sharp contrast, visible as "rings" moving inland.

We validated the models by computing the root mean square error (RMSE) between the predicted air temperatures and the measured air temperatures at 30% of the traverse points (Table 4.4). In all cases, RMSE was lower using RF than using RTA and OLS. This result suggested that the RF model more accurately predicted the surface temperatures than the other methods for this study site. Ho et al. (2014) also found that the RA model produced the lowest prediction errors for mapping urban air temperatures. For the RF model, the difference between using all buffers and using selected buffers was negligible. Therefore, we used all buffer sizes as input variables rather than selecting the most influential buffer sizes for all variables.

We further assessed the important variables for predicting air temperature. Random forest produces a measure of the importance of the predictor variable, which is called a mean decrease in accuracy. For each tree, the prediction error on the out-of-bag portion of the data is recorded, and the same is done after permuting each predictor variable. The difference between the two is then averaged over all trees and normalized by the standard deviation of the differences (Liaw and Wiener 2002; Breiman and Cutler 2012). Table 4.5 lists the top three variables in terms of variable importance, and its value of the mean decreases in accuracy; a plus or minus sign shows whether the variable is positively or negatively related to temperature.

Across the 5 days of our sample, we noted three important results. First, regardless of day or time, the distance to the coast is the most important predictor of temperature.

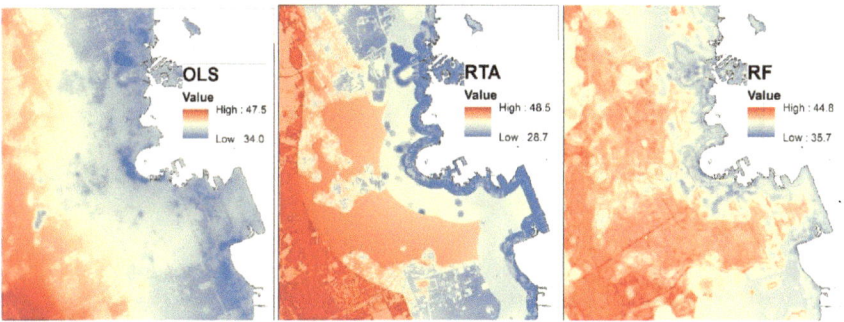

**Fig. 4.4** Predicted air temperature map (unit is in °C) using OLS (left); regression tree (center); and random forest (right) for 12 May 2015, 1 pm

**Table 4.4** Root mean square error (RMSE) between predicted near-surface air temperatures and measured temperatures at the testing site (in Celsius)

| | September 8, 2014 | | | September 9, 2014 | | |
|---|---|---|---|---|---|---|
| | 6 am | 1 pm | 7 pm | 6 am | 1 pm | 7 pm |
| OLS | 1.28 | 1.21 | 0.62 | 0.96 | 1.22 | 0.63 |
| RTA | 1.05 | 1.09 | 0.53 | 0.80 | 1.06 | 0.57 |
| RF | 0.43 | 0.75 | 0.34 | 0.35 | 0.79 | 0.39 |

| | May 12, 2014 | | | May 14, 2014 | | | May 15, 2014 | | |
|---|---|---|---|---|---|---|---|---|---|
| | 6 am | 1 pm | 7 pm | 6 am | 1 pm | 7 pm | 6 am | 1 pm | 7 pm |
| OLS | 1.19 | 1.81 | 1.72 | 1.44 | 2.71 | 0.96 | 0.79 | 1.15 | 1.00 |
| RTA | 0.93 | 1.44 | 0.82 | 1.34 | 1.33 | 0.76 | 0.72 | 1.05 | 0.93 |
| RF | 0.63 | 1.05 | 0.54 | 0.82 | 0.99 | 0.47 | 0.46 | 0.97 | 0.79 |

**Table 4.5** Top three variables in terms of variable importance derived from RF. Each variable is described as follows: w_dist: distance to the coast, a: albedo, v: vegetation, u: urban, the number indicates the buffer distance

|  | September 8, 2014 | | September 9, 2014 | | May 12, 2015 | | May 14, 2015 | | May 15, 2015 | |
|---|---|---|---|---|---|---|---|---|---|---|
| 6 am | w_dist | 91.9 | w_dist | 104.6 | w_dist | 75.7 | w_dist | 61.0 | w_dist | 67.0 |
|  | a50 | 47.5 | v600 | 52.1 | u600 | 32.0 | a50 | 35.4 | v600 | 31.8 |
|  | v600 | 41.1 | v550 | 38.3 | v600 | 28.5 | a600 | 31.0 | a600 | 25.7 |
| 1 pm | w_dist | 133.8 | w_dist | 134.0 | w_dist | 151.8 | w_dist | 63.4 | w_dist | 113.7 |
|  | v600 | 36.8 | u200 | 34.9 | a50 | 37.6 | v550 | 19.6 | a50 | 26.5 |
|  | a150 | 31.0 | u350 | 34.6 | a600 | 29.8 | a600 | 19.3 | v600 | 21.7 |
| 7 pm | w_dist | 105.8 | w_dist | 129.2 | w_dist | 59.4 | w_dist | 105.9 | w_dist | 73.5 |
|  | v600 | 37.3 | v600 | 37.5 | v600 | 24.8 | v550 | 23.6 | v600 | 24.1 |
|  | a50 | 37.1 | a50 | 35.6 | v550 | 20.2 | v600 | 19.4 | v550 | 23.9 |

Specifically, in the afternoon (1 pm), the large value of the distance to the coast has its most significant impact on temperature. This is a result of the relative temperatures of the sea surface and land surface. Specifically, while daytime air temperatures on the land are in the range of 33–48 °C, the average sea surface temperature in the gulf during the summer is 33 °C (World sea temperature. Doha average August sea temperature 2016). This temperature difference drives a sea breeze that provides local cooling near the coast. Although the prevailing wind in most months is from the North–Northwest in Doha, these winds are from the East and Northeast in summer. We acquired wind direction data from the weather station near Doha's city center, and the data supported this trend. Second, for 5 days in the evening (7 pm), the second most important variable was the measures of vegetation. Third, the results for the morning indicated the most inconsistency across 5 days for predicting the important variable. For example, the morning results indicated that the proximity to the water had both a cooling and heating effect, depending on the day. Overall, the results suggest that higher albedo corresponds to lower temperatures in the morning and evening, while farther distances from the coast result in higher temperatures in the afternoon and evening. Finally, more vegetation corresponds to lower temperatures in the evening.

Based on these results, we used the random forest model to divide the region into five categories, from the lowest (T1) to the highest temperatures (T5). By combining the time periods, we were able to identify those areas exhibiting the hottest temperatures (Fig. 4.5). A visual description of the hottest areas in the morning shows that they are near the central business district and at the Doha international airport, which is located near the southeast side of the coast. In the afternoon, the hottest areas are located furthest from the coast. In the evening, the hottest areas are located along two major highways, Salwa Road and Doha Expressway.

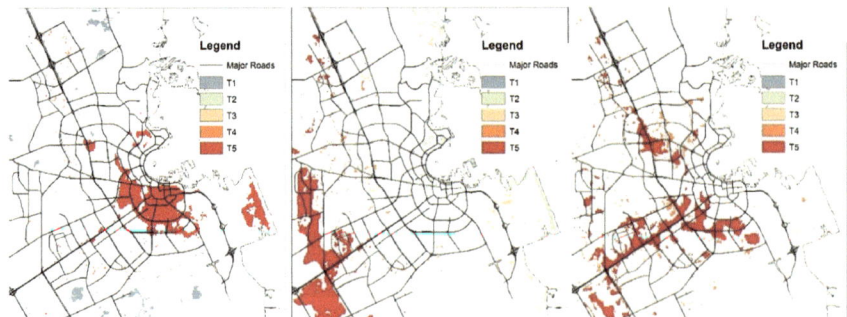

**Fig. 4.5** Air temperature variability maps with road maps for 5 days in each time period: 6 am (left); 1 pm (center); 7 pm (right)

## 4.8  Implications for Managing Regional Air Temperatures

The creation of impervious surfaces is centrally important to the creation of cities. In areas where high temperatures can cause major health impacts, understanding the role of landscape features is essential to developing mitigation strategies. Our analysis provides insight into the role of local land cover and times of day in temperature variations across one city. Unlike earlier studies that suggest one urban heat island for a city, we were able to illustrate a dynamic description of urban heat effects throughout a day. Our empirical assessments, models, and predictions indicated that urban heat islands migrate throughout the day. This is the first such description of the phenomenon that we know of. We speculate that specific features of the landscape gain thermal capacity at different rates and also release temperatures at different rates. This results, our analysis suggests, in locations where such accumulations and dissipations of thermal difference occur.

These results suggest specific recommendations for future development in the region. While temperatures during the middle of the day may be intolerable for the human body, our results suggest that the interaction between land-cover and temporal variation may lead to opportunities for reducing temperatures during "shoulder periods," which we define as transition times during late morning and early evenings, and may offer chances for changing the urban design such that more people can spend time outside. If planning agencies are considering options for mediating temperatures to provide pedestrians greater access to outdoor spaces, then reducing the amount of urban area (a direct measure of impervious surfaces) may be the first step. While changing land cover may not be cost-effective or a feasible option in places containing a large amount of impervious surface, covering the concrete with trees may be a reasonable alternative. Despite the arid climate, given the abundant amount of water from desalination in Doha, water resources for expanding the urban canopy may be readily available. Of course, this solution raises other questions of sustainability related to desalination effects on the salinity of gulf waters. Ultimately, there will be tradeoffs between heat mitigation strategies and other environmental factors that must be thoroughly considered.

As expected, we found that the albedo was negatively related to local temperatures in the morning and evening. However, this relationship was positive during midday periods. Although increasing the albedo of surfaces is a common practice, one that could reduce the absorption of solar radiation, it might not work for Doha. Higher albedo could help the temperature in the morning and evening, but it could also have the adverse effect of increasing temperatures during the daytime. Finally, the distance from the coastline indicates that the mediating influence of coastal waters can significantly impact inland air temperatures. If, however, coastal winds are blocked by high-rise buildings along the coastline, inland areas may not benefit from this mediating influence. Restricting development along the coast, especially those buildings that prevent these coastal processes from meditating inland temperatures, is a policy that has traction in scholarly research (Wong et al. 2011), and it may be a policy option that can improve the short- and long-term quality of life of Doha's residents.

In this study, several methodological limitations may obscure the implications of our findings. First, our study was designed as correlational and not causational. Therefore, we were only able to describe, in likelihood terms, the impact of different land-cover factors on air temperatures. Second, the maximum buffer size was set to 600 m, which was selected based on the existing literature (Hart and Sailor 2009) and our interest in identifying localized actions that could mitigate urban heat. The plot of the linear regression correlation coefficient between the buffer size and temperature for several variables, especially for the urban class, did not reach a plateau by 600 m. This indicated the possibility of a larger influential area. Reviewers have noted other studies indicating that land-cover changes at distances 1 km from the monitoring site could impact air temperatures, and we expected to examine those relationships in future research. Finally, one of our research questions asked what land-use and land-cover factors best describe the variation of urban heat throughout the day. Since land-use variables are generally not available to researchers in our study location, we only employed land-cover factors. Accordingly, we were not able to discern the extent to which the urban form of different land uses (e.g., commercial, single family, residential, etc.) affects urban heat. Finally, as Nichol (2005) found in a related study, satellite-derived surface temperatures are highly correlated to air temperatures. Our study did not include satellite-derived surface temperatures, and we believe that the accuracy of our models could be improved by adding relevant surface temperatures variables.

In future directions of this project, we expect to include satellite-derived surface temperatures, which will help us to understand the extent to which our techniques for assessing ambient temperatures can be enhanced through remotely sensed satellite data. Such an approach will dramatically reduce the costs of conducting urban heat analysis for cities across the planet. Further, we were unable to evaluate the effectiveness of varying land-cover modifications on ambient temperatures. In future work, we expect to expand these analyses by incorporating real and hypothesized changes into surface characteristics and materials. An emerging and promising literature suggests the need to provide context-specific applications for reducing urban heat, and we anticipate doing so in the near future.

The eventual outcome of the present project was to support decision-making efforts in order to improve livability in the ever-increasing outdoor temperatures of Doha. Several recent studies indicate that Doha and many other parts of the Middle East will encounter steady increases in daytime temperatures over the coming decades. Accordingly, our findings provide early evidence about the land-cover factors that can hinder or amplify outdoor air temperatures. Difficult to find is information about how the local government is evaluating current mitigation options. Still needed are additional studies that attempt to integrate our findings of the physical characteristic of Doha with decision-making frameworks that may reduce the harm caused by future increases in urban climates.

# References

Anderson JR, Hardy EE, Roach JT, Witmer RE (1976) A land use and land cover classification system for use with remote sensor data (US Geological Survey, Professional Paper 964)

Balbus JM, Malina CM (2009) Identifying vulnerable subpopulations for climate change health effects in the United States. J Occup Environ Med 51(1):33–37. https://doi.org/10.1097/JOM.0b013e318193e12e

Brazel A, Gober P, Lee S-J, Clarke-Grossman S, Zehnder J, Hedquist B, Comparri E (2007) Determinants of changes in the regional urban heat island in metropolitan Pheonix (Arizone, USA) between 1990 and 2004. Climate Res 33:12. https://doi.org/10.3354/cr033171

Breiman L (2001) Random forests. Mach Learn 45(1):5–32. Kluwer Academic Publishers. https://doi.org/10.1023/a:1010933404324

Breiman L, Cutler A (2012) Breiman and Cutler's random forests for classification and regression. In: Package 'randomForest', p 29. https://doi.org/10.5244/c.22.54

Brower BV, Laben CA (2000) U.S. Patent No. 6,011,875

Frey CM, Rigo G, Parlow E (2007) Urban radiation balance of two coastal cities in a hot and dry environment. Int J Remote Sens 28(789273183):2695–2712. https://doi.org/10.1080/01431160600993389

Hansen MC, DeFries RS, Townshend JRG, Sohlberg R, Dimiceli C, Carroll M (2002) Towards an operational MODIS continuous field of percent tree cover algorithm: examples using AVHRR and MODIS data. Remote Sens Environ 83:303–319

Hart MA, Sailor DJ (2009) Quantifying the influence of land-use and surface characteristics on spatial variability in the urban heat island. Theoret Appl Climatol 95(3):397–406. https://doi.org/10.1007/s00704-008-0017-5

Heusinkveld BG, Steeneveld GJ, Hove LV, Jacobs CMJ, Holtslag AAM (2014) Spatial variability of the Rotterdam urban heat island as influenced by urban land use. J Geophys Res-Atmos 119:677–692

Ho HC, Knudby A, Sirovyak P, Xu Y, Hodul M, Henderson SB (2014) Mapping maximum urban air temperature on hot summer days. Remote Sens Environ 154:38–45

Ho HC, Knudby A, Xu Y, Hodul M, Aminipouri M (2016) A comparison of urban heat islands mapped using skin temperature, air temperature, and apparent temperature (Humidex), for the greater Vancouver area. Sci Total Environ 544:929–938. https://doi.org/10.1016/j.scitotenv.2015.12.021

Jensen JR (1986) Introductory digital image processing. Prentice-Hall, Englewood Cliffs, NJ, USA

Knowlton K, Lynn B, Goldberg RA, Rosenzweig C, Hogrefe C, Rosenthal JK, Kinney PL (2007) Projecting heat-related mortality impacts under a changing climate in the New York City region. Am J Public Health 97(11):2028–2034. https://doi.org/10.2105/AJPH.2006.102947

Kohavi R (1995) A study of cross-validation and bootstrap for accuracy estimation and model selection. IJCAI 14:1137–1145

Krüger E, Givoni B (2007) Outdoor measurements and temperature comparisons of seven monitoring stations: preliminary studies in Curitiba, Brazil. Build Environ 42(4):1685–1698. https://doi.org/10.1016/j.buildenv.2006.02.019

Lazzarini M, Marpu PR, Ghedira H (2013) Temperature-land cover interactions: the inversion of urban heat island phenomenon in desert city areas. Remote Sens Environ 130:136–152. https://doi.org/10.1016/j.rse.2012.11.007

Li J, Song C, Cao L, Zhu F, Meng X, Wu J (2011) Impacts of landscape structure on surface urban heat islands: a case study of Shanghai, China. Remote Sens Environ 115(12):3249–3263. https://doi.org/10.1016/j.rse.2011.07.008

Liaw A, Wiener M (2002) Classification and regression by randomForest. In: R News, pp 18–22

Lillesand T, Kiefer RW, Chipman J (2014) Remote sensing and image interpretation. Wiley, New York, NY, USA

Lo CP, Choi J (2004) A hybrid approach to urban land use/cover mapping using Landsat 7 Enhanced Thematic Mapper Plus (ETM+) images. Int J Remote Sens 25(14):2687–2700

Luber G, McGeehin M (2008) Climate change and extreme heat events. Am J Public Health 35:429–435

Nassar AK, Alan Blackburn G, Duncan Whyatt J (2014) Developing the desert: the pace and process of urban growth in Dubai. Comput Environ Urban Syst 45:50–62. https://doi.org/10.1016/j.compenvurbsys.2014.02.005

Nichol J (2005) Remote sensing of urban heat islands by day and night. Photogramm Eng Remote Sens 71

Oke TR (1976) The distinction between canopy and boundary layer urban heat islands. Atmosphere 14:268–277

Oke TR (1995) The heat island of the urban boundary layer: characteristics, causes and effects. Wind climate in cities. Springer, Dordrecht, The Netherlands, pp 81–107

Prasad AM, Iverson LR, Liaw A (2006) Newer classification and regression tree techniques: bagging and random forests for ecological prediction. Ecosystems 9(2):181–199. Springer. https://doi.org/10.1007/s10021-005-0054-1

R Development Core Team (no date) R: a language and environment for statistical computing. The R Foundation for Statistical Computing, Vienna, Austria. http://www.r-project.org/. Accessed 20 April 2016

Rao PK (1972) Remote sensing of urban heat islands from an environmental satellite. Bull Am Meteor Soc 53:647

Rasul A, Balzter H, Smith C (2015) Spatial variation of the daytime surface urban cool island during the dry season in Erbil, Iraqi Kurdistan, from Landsat 8. Urban Clim 14:176–186

Reid CE, O'Neill MS, Gronlund CJ, Brines SJ, Brown DG, Diez-Roux AV, Schwartz J (2009) Mapping community determinants of heat vulnerability. Environ Health Perspect 117:1730

Santamouris M, Ding L, Fiorito F, Oldfield P, Osmond P, Paolini, R, et al (2017) Passive and active cooling for the outdoor built environment—Analysis and assessment of the cooling potential of mitigation technologies using performance data from 220 large scale projects. Solar Energy 154:14–33

Schowengerdt RA (1997) Remote sensing: models and methods for image processing, (No. 621.367 S476 1997). Academic Press, Boston, MA, USA

Shandas V, Makido Y, Hong C, Ferwati S, Sailor D (2016) Rapid urban growth and development patterns in the Middle East: the case of Doha, Qatar. Unpublished work

Smith RB (2010) The heat budget of the Earth's surface deduced from space. Available online

Su YF, Foody GM, Cheng KS (2012) Spatial non-stationarity in the relationships between land cover and surface temperature in an urban heat island and its impacts on thermally sensitive populations. Landsc Urban Plan 107:172–180

Therneau T, Atkinson B, Ripley B, Ripley MB (no date) Package 'rpart'

Voogt JA, Oke TR (2003) Thermal remote sensing of urban climates. Remote Sens Environ 86(3):370–384

Wong MS, Nichol J, Ng E (2011) A study of the "wall effect" caused by proliferation of high-rise buildings using GIS techniques. Landsc Urban Plan 102(4):245–253. https://doi.org/10.1016/j.landurbplan.2011.05.003

World sea temperature. Doha average August sea temperature (2016)

Yale Center for Earth Observation (no date) How to convert landsat DNs to Albedo. http://yceo.yale.edu/how-convert-landsat-dns-albedo. Accessed 20 April 2016

Yan H, Fan S, Guo C, Wu F, Zhang N, Dong L (2014) Assessing the effects of landscape design parameters on intra-urban air temperature variability: the case of Beijing China. Build Environ 76:44–53. https://doi.org/10.1016/j.buildenv.2014.03.007

Yokobori T, Ohta S (2016) Effect of land cover on air temperatures involved in the development of an intra-urban heat island. Climate Res 39:61

Yuan F, Bauer ME (2007) Comparison of impervious surface area and normalized difference vegetation index as indicators of surface urban heat island effects in Landsat imagery. Remote Sens Environ 106(3):375–386. https://doi.org/10.1016/j.rse.2006.09.003

Yuan F, Wu C, Bauer ME (2008) Comparison of spectral analysis techniques for impervious surface estimation using Landsat imagery. Photogramm Eng Remote Sens 74:1045–1055

Zha Y, Gao J, Ni S (2003) Use of normalized difference built-up index in automatically mapping urban areas from TM imagery. Int J Remote Sens 24:583–594

Zhou W, Qian Y, Li X, Li W, Han L (2014) Relationships between land cover and the surface urban heat island: seasonal variability and effects of spatial and thematic resolution of land cover data on predicting land surface temperatures. Landsc Ecol 29:153–167

# Chapter 5
# Urban Form and Variation in Temperatures

Cynthia Skelhorn, Salim Ferwati, Vivek Shandas and Yasuyo Makido

**Abstract** This chapter examines microclimate modeling to evaluate the effects of changing landscape features on ambient temperatures in Doha, Qatar. By modeling three study sites around Doha—one highly urbanized, one newly urbanizing, and one coastal low-density urbanized—the research indicates that at the neighborhood-scale, the most effective scenario was that of adding mature trees along the sides of roads. In the coastal study area, the model results estimated a maximum hourly air temperature reduction of 1.35 °C, and in the highly urbanized inland site, surface temperature reductions were up to 15 °C at 12:00. While other scenarios were effective at reducing air and surface temperatures, the mean radiant temperature was also increased or nearly neutral for most of the other scenarios. This result highlights the need to develop improved shading measures for pedestrian pathways and outdoor recreational areas, especially for highly urbanized inland areas in Doha and cities with similar climatic conditions.

**Keywords** Extreme heat · Doha · Corridor · Scenarios · Mitigation effects

Recent evidence suggests that many densely populated areas of the world will be uninhabitable in the coming century due to depletion of resources, climate change, and increasing urbanization leading to changes in microclimate which directly affect livability (Lelieveld et al. 2016; Pal and Eltahir 2016). An increasing number of cities already experience temperature and humidity ranges that challenge the ability of residents to find acceptable levels of comfort outdoors during summer and even the hotter periods of the day in milder seasons, such as spring and autumn. A recent study of future temperatures in Southwest Asia found that many cities in the region of the Arabian Gulf will become uninhabitable by 2100 under current climate change

---

Sections of this chapter are from the following document, which is part of the Creative Commons Open Licensing system: Ferwati, S., C Skelhorn, V Shandas, and Y Makido, in 2019. A Comparison of Neighborhood-Scale Interventions to Alleviate Urban Heat in Doha, Qatar. Sustainability 11(3), 730.

projections (Pal and Eltahir 2016) while a Lancet article details the population–environment–development dynamics that require an urgent focus on survivability for the Arab world (El-Zein et al. 2014). The potential for some populous regions of the world to become uninhabitable poses serious questions about actions that require immediate attention and opportunities to stave off massive human population health impacts.

This chapter examines the role of specific landscape features as they affect ambient temperatures in one of the fastest growing regions of the world: Doha, Qatar. The city of Doha offers several advantages to examining the application of heat reducing strategies, including the fact that the average minimum and maximum temperatures range from 13 to 42 °C (from cool to extreme hot), and can exceed 50 °C as an hourly maximum during the summer. Additionally, recent research about Doha's urban climate suggests that the land-use and land-cover change are likely increasing ambient temperatures (Makido et al. 2016; Ferwati et al. 2018), which further amplifies emerging livability challenges in the region (AlSarmi and Washington 2014; Rizzo 2014). We build on an emerging literature about thermal comfort and livability in the Middle East by asking three questions: (1) to what extent are conventional treatments for cooling neighborhoods effective in an arid desert climate? (2) what strategies show the most promise for cooling the ambient temperature? and (3) how do similar cooling treatments vary in their effectiveness across diverse land-use and land-cover conditions in the city? We address these questions using empirical data, which were collected over the course of three years, and the application of state-of-the-art fluid dynamic modeling systems. To contextualize the study, we begin by fleshing out a few of the main challenges facing the region, and then describe our methods for addressing these questions.

With the extremely hot temperatures in Qatar generally, and Doha specifically, planners, engineers, and architects must consider the potential impacts of climate change. The Lancet, Britain's premier health journal, calls climate change "the biggest global health threat of the 21st century" (Costello et al. 2009). Qatar is predicted to have 65 days of heat wave conditions per year by 2020 under the RCP8.5 (business-as-usual) modeling scenario, with this number expected to increase to 88 by 2050 (ESRI 2017). Exposure to hot days may result in heatstroke(or, if directly in the sun, it is referred to as sunstroke), a life-threatening condition when a person collapses suddenly due to overheating (Das 2017). A number of other direct and indirect impacts have been studied, including increased risk of cardiovascular disease, worsening of respiratory conditions, risk of decreased kidney function, risk of adverse birth outcomes, and the changes in the distribution of vector-borne disease (Watts et al. 2018). Also, reduced labor productivity is already being seen in Gulf countries and elsewhere, which has a direct impact on both economic and well-being outcomes.

Some of the urban heat can be tempered through urban design, yet as Qatar has grown, little attention was given to the implications of climate change. As an arid peninsula projecting into the Gulf, Qatar has approximately 563 km of coastline and was primarily made up of small fishing and pearling settlements until recently. Since the late 1940s, Doha, the largest city and capital of Qatar, has experienced stages of urban growth ranging from traditional light-colored, low-rise, densely built neigh-borhoods to ambitiously sprawling low-density neighborhoods and a coastal center of high-rise towers through the implementation of small- to mega-scaled transfor-mative development projects. While much of Doha's built environment is designed like cities across the northern hemisphere, what has taken many centuries to evolve in other places has happened in less than half a century in Doha and often has been noted as lacking the integration between the master planning and implementation phases that might have otherwise led to an overall coherent city plan (Rizzo 2014). With increasing concerns about climate change impacts and urbanization for such an extreme desert climate as that of Qatar, government planners and environmental man-agers in Doha and other Qatari cities are examining strategies to modify its further growth with a shift to more sustainable and livable results (Qatar General Secretariat for Development Planning 2011; Qatar Ministry of Municipality and Environment 2016).

The existing built environment and projections of future changes to ambient tem-peratures in Doha pose several challenges for adapting to climate change. Climate models suggest that summer heat will be intensified by the UHI effect with recent research noting increasing trends of warm temperature extremes (AlSarmi and Wash-ington 2014). Furthermore, a study specific to climate change projections in the Gulf region predicts that, under a business-as-usual scenario of future greenhouse gas con-centrations, many areas of the GCC region are likely to become uninhabitable due to intolerable rises in the wet-bulb temperature (Pal and Eltahir 2016). For Doha, the combined effects of a highly built-out urban landscape, with expectation for increas-ing temperatures, and rapid urbanization create an acute need for rapid improvement in urban design practices.

We propose an examination of Doha's temperature, humidity, and thermal comfort conditions in neighborhoods containing varying land-use and land-cover conditions. Considering the distinction between the urban canopy layer (UCL), which is ground height to approximately roof height, and urban boundary layer (UBL) which extends above roof level (Oke 1976; Arnfield 2003), we examine the micro-scale relationships between land cover, building materials, and spatial variations in three study areas. The aim of the present study is to understand which physical features of urban design can improve thermal comfort in outdoor spaces during periods of intense heat. We propose that the intensity of the near-surface air temperature is correlated with the density and the distribution of certain impervious surfaces and materials and that selective change of these surfaces and materials can improve thermal conditions and improve thermal comfort.

## 5.1  Modeling Urban Form and Temperature

To address the research aims of the present study, we draw on several datasets and modeling system. The empirical datasets are derived from weather stations that the authors distributed throughout the study region. A total of ten weather stations were assembled and stationed in areas representing diverse physical geographies and built environments. The weather stations use solar power to track temperature, humidity, and wind speed. Tethered to an Omega datalogger, each weather station samples each parameter at 10-minute increments.

The weather station data were instrumental in calibrating the modeling system, ENVI-met. ENVI-met™ (V4.1.3, Winter Release 2016–2017) is a complex fluid dynamic model that uses a series of input parameters for each site to provide a spatially explicit description of the distribution of microclimates. The model has been developed and regularly improved since 1998 (Bruse and Fleer 1998; Bruse 2017) and was selected because it is specifically designed for investigating changes to the landscape and the built environment in urban areas.

The model uses a computational fluid dynamics (CFD) approach, employing the Reynolds-averaged non-hydrostatic Navier–Stokes equations for the wind field, the k-epsilon turbulence model and a combined advection–diffusion equation with the alternating directly implicit (ADI) solution technique (Bruse 2011) to model the interaction between microclimate and urban surfaces, such as walls, pavements, and vegetation. It employs both a soil model, with soil temperature calculated for natural soils and for artificial seal materials down to a depth of -4 m, and a vegetation model, which allows complex 3D vegetation geometries and accounts for heat and vapor exchanges with the atmosphere. ENVI-met uses the finite difference method to solve the numerous partial differential equations (PDE) (Bruse 2017). Spatial resolution can be selected in the range of 0.5–10 m. While the model is three dimensional, our model components are currently placed on a flat surface terrain, which is representative of terrain in Qatar.

ENVI-met requires two main input files: a configuration file, which contains settings for initialization values and timings; and an area input file, which allows the user to design the layout of the site to be modeled, including the location and types of buildings, trees and other vegetation, surfaces (e.g., asphalt and concrete), and soils. Areas to be modeled are digitized on a rectangular grid with the vertical top of the model set at a maximum of 2500 m, and a surface area of up to $250 \times 250$ cells). Data are input on a cell-by-cell basis. Points that are of interest for a careful examination (e.g., for corresponding field measurements) can be specified as receptor points. Timing of model runs is set by the user, with a typical model simulation starting from 6 am, simulating a period of 12–24 h and saving data once per hour (or more frequently, as necessary). Due to the extensive calculations and numerous outputs, the time required to complete one simulation period can be 24–48 h or more depending on the size of the model area, spatial resolution, and total number of hours simulated (Skelhorn et al. 2014).

## 5.2   Site Selection

The overall study of climate adaptation within which this piece of research was performed had installed a total of nine weather stations in the vicinity of the capital city

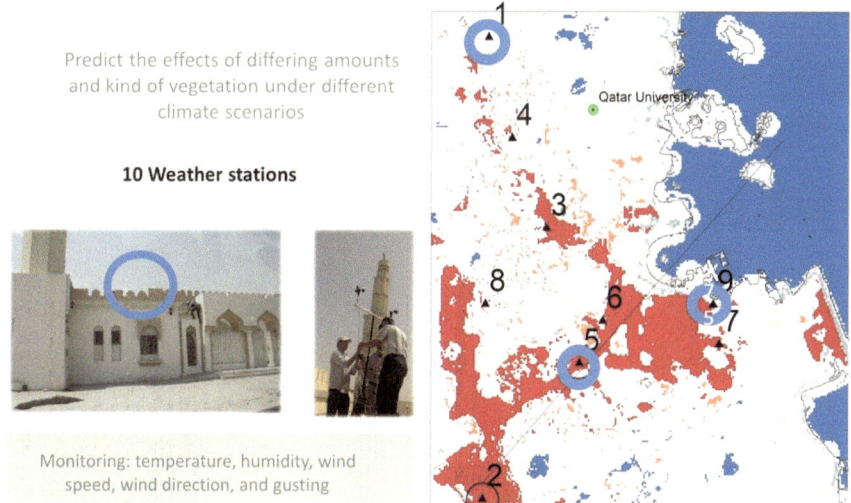

**Fig. 5.1** Locations of three case study areas, selected from neighborhoods in the vicinity of ten weather stations

of Doha, with access to the tenth station at Qatar University. We selected three neighborhoods from these weather station locations (Fig. 5.1) that were diverse in terms of their geographic distribution and land-use characteristics, but also representative of several styles of development currently found. The selected neighborhoods can be characterized as urbanized—the area surrounding weather station 5 in Al Waab; coastal—surrounding weather station 9 in Al Khulaifat; and developing—surrounding weather station 1 in Umm Salal Mohammad. Based on the availability of robust weather station data, the model was parameterized for a summer day in September 2014, using the urban microclimate model ENVI-met V4 (released July 1, 2016).

Since each site will undergo specific heat mitigation treatments, we note several features that require further explication. The developing site (site 1) in Umm Salal Mohammad was approximately one-third site developed at the time of the study, with a mosque in the center and one compound of two-storey villas in the northern corner of the site, with a few more compound villas along the southern edge of the site. The remainder of the site was bare soil.

The urbanized site (site 5) in Al Waab represents a dense existing urban development in Doha, with very little space for additional buildings, open space, trees, or other vegetation (Fig. 5.2). It consists of primarily two-storey residential buildings that are approximately 6–8 m in height, all behind a walled perimeter with many areas containing dense mature trees that are 10–12 m in height. The southeastern edge of the area is bordered mainly by covered parking structures that provide additional parking for nearby businesses and residents. All streets are two-lane residential roads with additional parking lanes on both sides. This site represents the typical building pattern in residential areas over the past 20–30 years.

**Fig. 5.2**   Photo survey of Site 5

**Fig. 5.3**   Photo survey of Site 9

Finally, the coastal site (site 9) in Al Khulaifat represents a site with potential for redevelopment (Fig. 5.3). The southern half of the site has a parking area, a mosque, and a small private group of low buildings (approximately 6 m in height), and an area of very well-landscaped grounds in the southwest quadrant. The remainder of the site consists of the main road running east to west about halfway from north to south on the site, while the rest (northern half) is bare soil.

## 5.3 Model Setup and Calibration

Each of the three selected sites was digitized in ENVI-met using a color-coded background image. The input configurations for the sites are shown in Fig. 5.4.

The models were calibrated based on measured air temperature data from the weather station located on each site and then analyzed for microclimate changes due to each change in material or vegetation. The measurement data served two purposes. First, the weather station provided empirical data for comparing air temperatures in different locations around Doha. Second, the data allowed calibration of the ENVI-met model by setting a receptor in each model that corresponded to the weather station location. This approach aided in validation of the modeling results. Model calibration is important when working with a model such as ENVI-met, as results can vary widely if the model is not calibrated to site-specific conditions (Fig. 5.5). Appendix shows the changes in initial parameters for each version of the base model calibration. Once a reasonable base model was obtained, one change was made in each scenario and then results were compared to the base model. For this study, the hours of 6 am to 6 pm were chosen as this is the hottest part of the day and the time

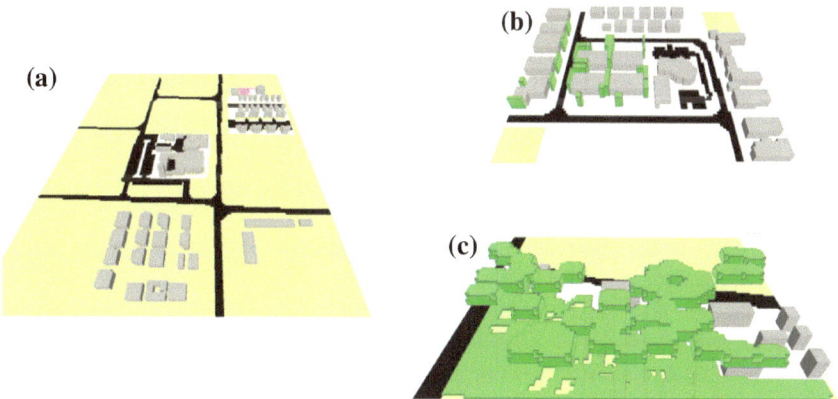

**Fig. 5.4** Color-coded representation of study areas in ENVI-met, **a** Site 1 (Umm Salal Mohammad), **b** Site 5 (Al Waab), **c** Site 9 (coastal) (gray = buildings, black = asphalt, yellow = sandy soil, green = vegetation)

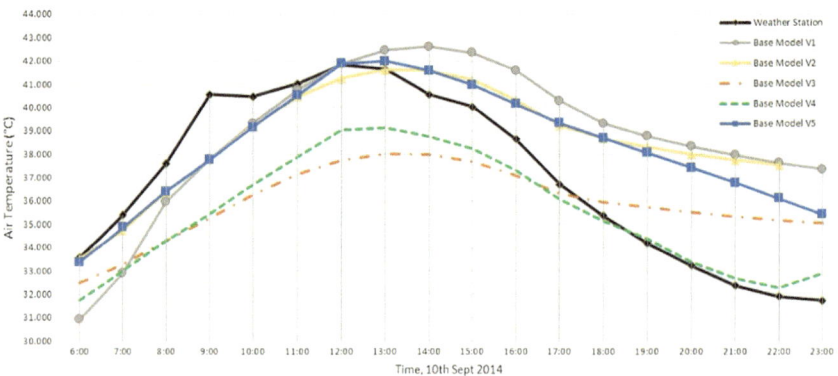

**Fig. 5.5**  Calibration of Site 1 base model, comparison of model results to weather station data for version 1 to version 5 (see Appendix for model settings)

frame for which the design improvements are likely to have the most impact on the comfort of residents.

## 5.4  Modeling Scenarios

Following the setup of a base model for each site, which represents the current field conditions, several additional scenarios were tested in which one substantial change was made either to the materials, the vegetation, or the configuration of buildings. Various studies have advanced several key urban heat mitigation strategies, primarily, (1) urban forestry, (2) cool roofs, and (3) cool pavements (McPherson et al. 1994; Golden 2004; Santamouris 2013, 2014; Kyriakodis and Santamouris 2018). Recently, blue space strategies (e.g., ponds, waterfronts, rain gardens, and water features) (Hathway and Sharples 2012; Gunawardena et al. 2017) and building arrangements (Ali-Toudert and Mayer 2006; Hong and Lin 2015; Deng et al. 2016) have also been investigated. In this study, we chose to assess realistic changes in building and landscape that could cool the urban environment in Qatar. Given that most buildings, especially residential, are already constructed of light-colored materials, including the roof, and that the arid environment is a somewhat limiting factor for water-based strategies, the remaining options to investigate include urban greening, changes to pavements, and changes to building arrangements. The model scenarios tested are described in Fig. 5.6.

While the range of variables output from ENVI-met is extensive, three key variables were selected to compare for this study. Studies of the UHI typically investigate changes in air temperature, surface temperature, or both, so these were initially selected. Mean radiant temperature ($T_{mrt}$) is typically considered as one of the most important variables related to thermal comfort under sunny conditions and is defined as the uniform temperature of a hypothetical spherical surface surrounding the sub-

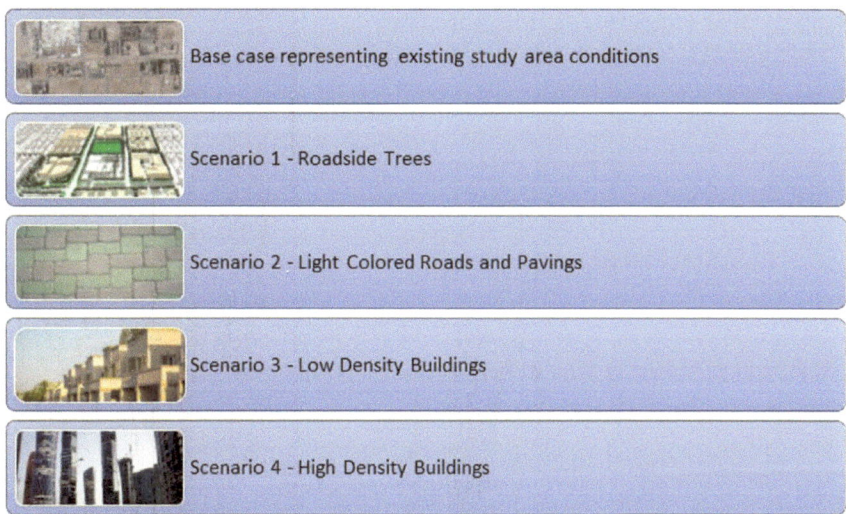

**Fig. 5.6** Scenarios tested in ENVI-met models (Low-density photo credit: https://cdn. forbesmiddleeast.com/wp-content/uploads/sites/3/2016/10/ryad0002.jpg; High-density photo credit: http://whyqatar.me/images/tornado_tower__palm_towers.jpg; other images by authors)

ject (emissivity $\varepsilon = 1$) that would result in the same net radiation energy exchange with the subject as the actual, complex radiative environment (Matzarakis et al. 2007; Walikewitz et al. 2015). It is a required parameter in the calculation of several other thermal comfort indices, including PET (Physiologically Equivalent Temperature), the PMV (Predicted Mean Vote), UTCI (Universal Thermal Climate Index), and the PT (Perceived Temperature) (Blazejczyk et al. 2012). In this study, we specifically wanted to compare both the effects of the selected changes on the UHI and on thermal comfort.

## 5.5 Comparing Urban Forms

The three base models were tested using a realistic set of input parameters until a base model that could reasonably approximate the measured air temperatures was developed. This section presents a comparison of the base model results to show overall thermal patterns on each site for the modeled day (10th September 2014).

## 5.6 Air Temperature

Table 5.1 presents the results of the daytime base model temperatures for the three sites at 6 am, 12 pm, 2 pm, and 6 pm. In general, we find that the urbanized site (5) has the lowest temperatures during the earlier parts of the day. However, this pattern changes after midday and by 2 pm, it is the warmest (with a maximum of 46 °C) and

**Table 5.1** Base model daytime air temperatures (°C)

|  | Site 1—Umm Salal Mohammed (Developing) | Site 5—Al Waab (Urbanized) | Site 9—Al Khulaifat (Coastal) |
|---|---|---|---|
| *Base model results: Air temperature—6 am* | | | |
| Min | 32.44 | 29.81 | 31.89 |
| Max | 33.38 | 31.71 | 33.62 |
| Avg | 32.91 | 30.76 | 32.75 |
| *Base model results: Air temperature—12 pm* | | | |
| Min | 40.46 | 39.85 | 41.37 |
| Max | 42.61 | 42.93 | 43.75 |
| Avg | 41.53 | 41.39 | 42.56 |
| *Base model results: Air temperature—2 pm* | | | |
| Min | 41.92 | 42.21 | 42.32 |
| Max | 44.03 | 46.26 | 44.59 |
| Avg | 42.98 | 44.23 | 43.45 |
| *Base model results: Air temperature—6 pm* | | | |
| Min | 39.29 | 40.23 | 39.13 |
| Max | 39.40 | 41.19 | 39.46 |
| Avg | 39.51 | 40.71 | 39.30 |

remains so until 6 pm. The coastal site (9) shows similar minimum temperatures to the other two sites at 2 pm, but the maximum temperature is slightly higher than that of the developing site (1), and approximately 1.6 °C lower than that of the urbanized site. Overall, the daytime pattern of temperatures is similar for the developing (1) and coastal (9) sites, but with the coastal site showing the highest maximum temperatures in the early morning, as expected for a site near a large water body.

## 5.7   Surface Temperature

Focusing on the period of the day with most intense heat, the surface temperatures at 2 pm range from 33 to 63 °C, with Site 5 (Urbanized) showing the highest maximum (Table 5.2). This is not much higher than Site 1 (Developing) but is nearly 7 °C higher than Site 9 (Coastal). At that time of day, Site 5 is almost uniformly hot, at approximately 54 °C, with some areas that are approximately 47 °C between buildings and small pockets of distinctly cooler surfaces at the western edges of buildings in the middle of the study area. Site 9 (Coastal) has very slightly higher temperatures in the early morning but remains distinctly cooler than the other two sites throughout midday. The coastal site (9), which has mature trees on the southern half of the site, registers lower surface temperatures by approximately 2–3.5 °C on average during midday.

**Table 5.2** Base model daytime surface temperatures (°C)

| | Site 1—Umm Salal Mohammed (Developing) | Site 5—Al Waab (Urbanized) | Site 9—Al Khulaifat (Coastal) |
|---|---|---|---|
| *Base model results: Surface temperature—6 am* | | | |
| Min | 27.62 | 28.41 | 28.46 |
| Max | 32.85 | 32.85 | 33.19 |
| Avg | 30.24 | 30.63 | 30.83 |
| *Base model results: Surface temperature—12 pm* | | | |
| Min | 32.85 | 32.85 | 32.85 |
| Max | 60.39 | 60.00 | 55.88 |
| Avg | 46.62 | 46.43 | 44.37 |
| *Base model results: Surface temperature—2 pm* | | | |
| Min | 32.85 | 32.85 | 32.85 |
| Max | 62.01 | 63.13 | 56.27 |
| Avg | 47.43 | 47.99 | 44.56 |
| *Base model results: Surface temperature—6 pm* | | | |
| Min | 32.85 | 32.85 | 32.85 |
| Max | 44.62 | 47.35 | 44.92 |
| Avg | 38.73 | 40.10 | 39.89 |

## 5.8 Mean Radiant Temperature

While mean radiant temperatures (Tmrt) do not show much difference within each site, they do show interesting differences between the sites (Table 5.3). The coastal site has a somewhat higher MRT in the early morning, especially compared to the developing site, but has the lowest MRT at midday—7.7 °C cooler than the highly urbanized site and over 4 °C cooler than the developing site at 2 pm. Small areas, which have an MRT up to 20 °C lower than the rest of the site, are found on the East and Northeast sides of vegetation and buildings at 2 pm.

## 5.9 Scenario 1—Roadside Trees

After calibrating and analyzing the base model for each site, the first scenario tested was the addition of trees, where trees were added at approximately 10 m intervals along the existing roads. Maximum air temperature of 1.2–1.3 °C were found for the developing and coastal sites (1 and 9), while the highly urbanized site (5), which has little available space for additional tree planting showed almost no change (Fig. 5.7). However, the highly urbanized site showed large reductions in surface temperature at 12 pm and 2 pm (up to 15 °C), while the other two sites had 4–8 °C reductions during

**Table 5.3** Base model daytime mean radiant temperatures (°C)

| | Site 1—Umm Salal Mohammed (Developing) | Site 5—Al Waab (Urbanized) | Site 9—Al Khulaifat (Coastal) |
|---|---|---|---|
| *Base model results* $T_{mrt}$—*6 am* | | | |
| Min | 20.30 | 21.53 | 23.38 |
| Max | 25.16 | 27.45 | 27.53 |
| Avg | 22.73 | 24.49 | 25.46 |
| *Base model results* $T_{mrt}$—*12 pm* | | | |
| Min | 56.56 | 59.98 | 56.62 |
| Max | 77.67 | 81.11 | 73.54 |
| Avg | 67.11 | 70.54 | 65.08 |
| *Base model results* $T_{mrt}$—*2 pm* | | | |
| Min | 58.18 | 61.21 | 55.73 |
| Max | 82.77 | 86.72 | 76.90 |
| Avg | 70.48 | 73.96 | 66.31 |
| *Base model results* $T_{mrt}$—*6 pm* | | | |
| Min | 33.42 | 35.68 | 33.55 |
| Max | 34.94 c | 38.35 c | 35.44 |
| Avg | 34.18 c | 37.02 c | 34.49 |

these times. Some small areas were found to have elevated surface temperatures, typically on the Eastern side of trees in the late afternoon. This could be caused by trapping of outgoing longwave radiation under tree canopies later in the afternoon and evenings. Interestingly, MRT was reduced by up to 19.52 °C at 2 pm for the developing site (1) and by similar amounts for the other two sites. Although MRT shows a slight increase across sites in the early morning, these results overall highlight the role of trees as an effective strategy for improving thermal comfort during the day with little or no drawback for evening or early morning hours.

## 5.10  Scenario 2—Lighter Roads and Pavers

For the scenario investigating roads and paving materials (changing to light gray pavers and a light red asphalt), several observations can be made based on the results in Fig. 5.8. It is primarily the surface temperatures that are improved (approximately 11 °C lower at midday for the coastal Site 9 and 7.2 °C lower for the developing Site 1), while only small changes in air temperature (also 0.6–0.7 °C for sites the coastal and developing sites, and up to 2.9 C for the urbanized Site 5, as in the scenario with added trees) are found. Surprisingly, MRT increases on average for the urbanized and coastal sites (5 and 9) during the daytime, while only very small reductions are found for the developing site. While the urbanized site (5) shows

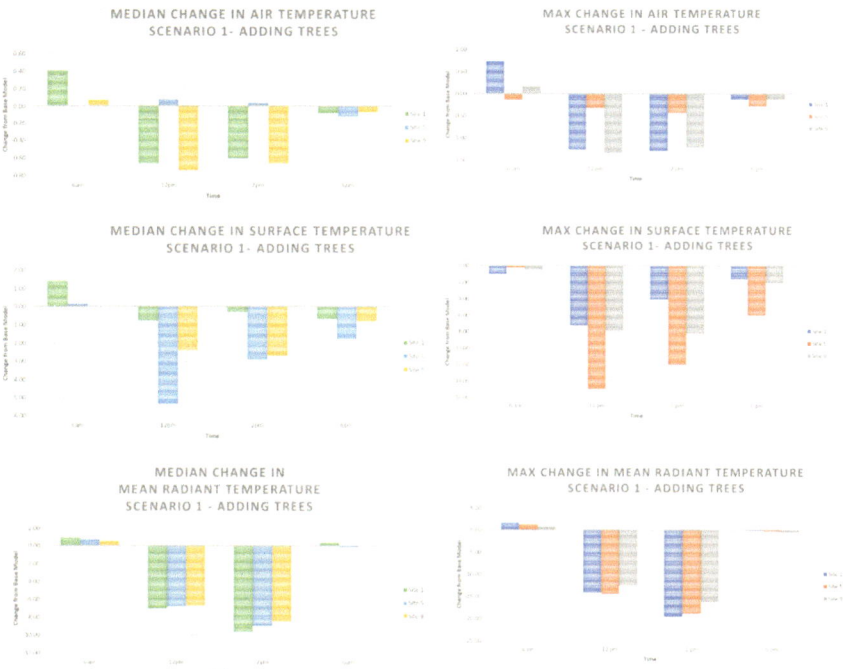

**Fig. 5.7** Median and maximum changes in air temperature, surface temperature, and mean radiant temperature for Scenario 1—Roadside Trees

the largest average and maximum reductions in air and surface temperatures, it also shows an average increase in $T_{mrt}$ through a change in roads and paving materials. As it is the site with the largest percentage of paved and asphalt surfaces, the finding is significant because demonstrates that, while near-surface air temperatures and surface temperatures may be reduced through a change in materials, the thermal comfort may be slightly worsened.

## 5.11   Scenario 3—Low-Density Buildings

To test the effects of differing building arrangements, we first tested a scenario of a typical low-density building arrangement that is common in Qatar and many other countries in the region. Buildings in this arrangement are usually 2–3 storey detached or semidetached, residences arranged in a rectangular layout inside a walled plot. One compound can contain a small number of villas (20–30), up to several hundred. As the developing and coastal sites (1 and 9) are largely undeveloped, the building scenarios were tested on these two sites. The highly urbanized site (5), due to its

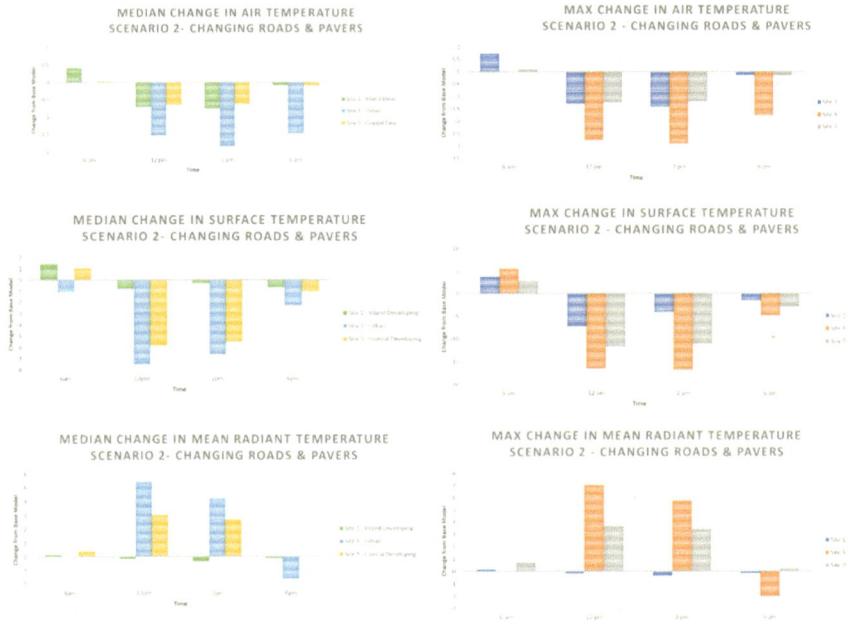

**Fig. 5.8** Median and maximum changes in air temperature, surface temperature, and mean radiant temperature for Scenario 2—Changing Roads and Pavers

currently dense building arrangement, was not modeled for any alternative building arrangements.

While the changes are quite similar between the two sites across all times of day, some notable differences can be found at midday (Fig. 5.9). Surface temperatures in early morning show higher maximum reductions for the coastal site (9), but this is mainly due to differences in small isolated patches at the north side of each building. At midday, reductions in both $T_{mrt}$ and surface temperature are found on both sites, but again, these are in small strips around the northern and eastern edges of the buildings, due to slight building overshadowing at that time of day. Other areas between the buildings and on the southern side show increases of approximately 2.6 °C for surface temperature and almost no change for $T_{mrt}$. While the patterns are similar in terms of locations of increases and reductions for both sites, the absolute change appears to be greater for the developing site (1) than for the coastal site (9). Interestingly, in the developing site, in which we tested several areas of the courtyard-style building arrangements, compared to semidetached rows of buildings, the interiors of courtyards show reductions of approximately 6 °C for surface temperature, while the semidetached rows showed increases of approximately 9 °C on roads. Air temperature reductions, typically between 1 and 1.5 °C, were more uniform across the site and depended on the localized air temperature.

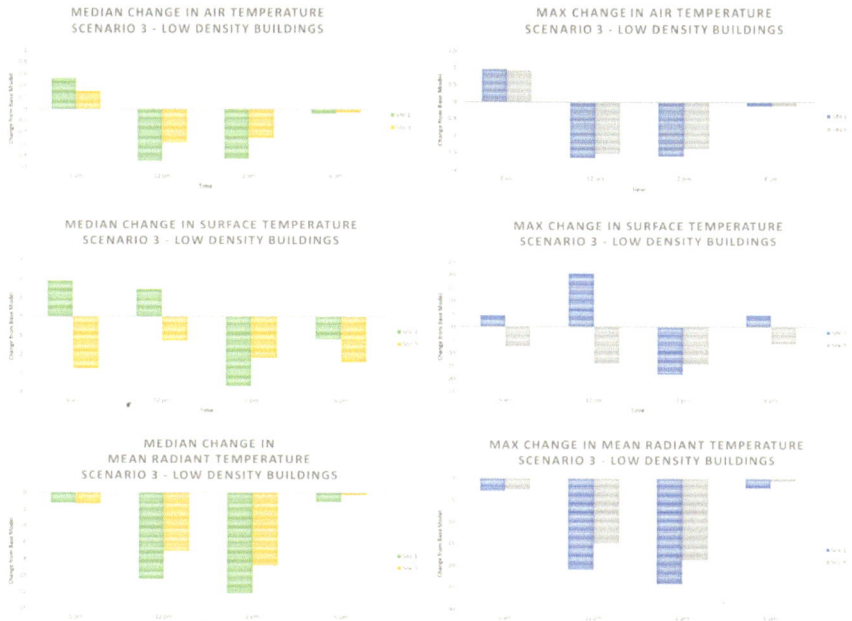

**Fig. 5.9** Median and maximum changes in air temperature, surface temperature, and mean radiant temperature for Scenario 3—Low-density buildings

## 5.12   Scenario 4—High-Density Buildings

For the high-density building scenario, we tested the sites with tower-block-style arrangements of square buildings of 30 m height. Both sites, developing and coastal, showed large reductions in all parameters, especially surface temperatures and $T_{mrt}$ during midday, with Site 1 showing the largest reductions at 12 pm and 2 pm (Fig. 5.10). However, it should be noted that these areas of large reductions are limited to small strips in the shadows of buildings, as in the scenario for the low-density building arrangements. All areas of paved surfaces (either asphalt or paving blocks) showed increases of surface temperature and $T_{mrt}$ of 4–8 °C at midday. The temporal patterns of temperature change are quite similar between low-density and high-density, with slightly greater reductions in air temperature and $T_{mrt}$ in the high-rise scenario at midday in the developing site (1) and for all three parameters in the coastal site (9).

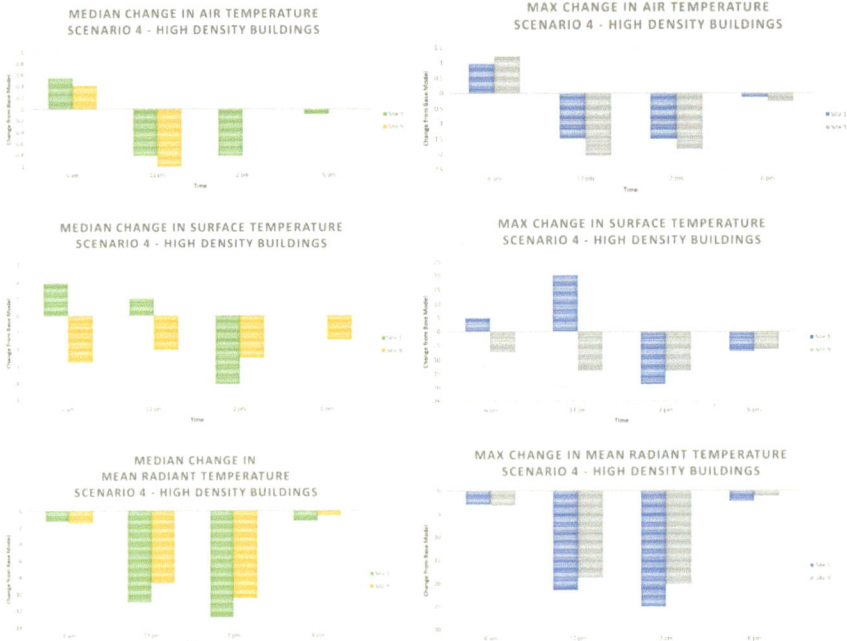

**Fig. 5.10** Median and maximum changes in air temperature, surface temperature, and mean radiant temperature for Scenario 4—High-density buildings

## 5.13  Implications for Urban Design

The four scenarios presented here provided comparisons of air temperature, surface temperature, and mean radiant temperature (an indicator of thermal comfort) between the base model and the models that have been changed with (1) additional tree cover along roadways; (2) higher albedo roads and pavements; (3) low-density building arrangements, and (4) high-density building arrangements. This section explores the most effective strategies related to the parameters investigated.

## 5.14  Air Temperature Reduction Strategies

For air temperature near the ground (1 m height), the model scenario with added trees is lower by approximately 0.6–0.7 °C, depending on the existing temperature regime in that area, for the developing and coastal sites (1 and 9). For the urbanized site, the air temperature reductions are very minimal, but this is reasonable considering that the site has fewer available areas for planting new trees and also that the change is from a paved/asphalt area to a shaded paved/asphalt area. For the developing and coastal sites, the change is from bare soil to shaded bare soil. In the highly urbanized Site

5, changing to lower albedo pavers and roads also demonstrates a good reduction in air temperature (up to 2.9 °C) at 2 pm. The developing and coastal sites show reductions of 1.2–1.4 °C maximum. For building changes, the high-density building scenario shows the strongest improvement for air temperature, but only in small areas in the afternoon (12 pm and 2 pm) shadows of buildings. For the developing site, slightly greater reductions in air temperature are found in the low-density building scenario. Other research in the temperate climate of the UK has found similarly modest reductions in air temperature due to shading, 0.1–0.2 °C for a 5% increase in tree cover across a suburban area (Skelhorn et al. 2014), but up to 0.9 °C for shaded areas compared to open areas of a park (Armson et al. 2012). A meta-analysis by Bowler et al. (2010) investigating temperature differences between parks and their urban surroundings estimated an average air temperature reduction of 0.94 °C by synthesizing results from a large number of studies from many different geographic locations and climatic regimes.

## 5.15  Surface Temperature Reduction Strategies

In examining the surface temperature differences, we find that the areas immediately under trees have surface temperatures that are reduced by up to 15 °C in the urbanized Site 5, while the other two sites show average reductions of 0.7–2.7 °C at midday. The urbanized Site 5 represents a change from pavers or asphalt to a shaded paved surface, whereas the developing and coastal sites represent a change from bare soil to shaded bare soil. Although the low-density and high-density buildings show large maximum reductions in surface temperatures, these are limited to small areas around the edges of buildings. Therefore, average surface temperature reductions across the whole site are greatest in the urbanized site, which has the largest amount of urbanized surface that is being changed to higher albedo materials.

While the scenario of changing the type of paver alone resulted in approximately 2 °C lower surface temperatures compared to the base model, we also found in comparing different areas of the base model that a standard light red paver will be lower by up to 12 °C than areas covered with asphalt and 3–4 °C cooler than areas of bare soil.

Only a few studies have attempted to analyze changes in surface temperature over time for different surface types. A recent study found that changing surfaces from concrete to grass reduced the surface temperature by up to 24 °C (Armson et al. 2012) in a UK site, whereas another study in Basel, Switzerland (Leuzinger et al. 2010), using a high-resolution thermal camera to take an aerial image, estimated that streets were approximately 11 °C warmer than vegetated areas. Using modeling studies, Kjelgren and Montague (1998) found that afternoon asphalt surface temperatures ($T_s$) were 20–25 °C higher than turf $T_s$, whereas Gill (2006) found a maximum difference of 25 °C between non-transpiring surfaces such as concrete and wholly transpiring surfaces such as woodland across Greater Manchester, UK.

## 5.16   Mean Radiant Temperature Reduction Strategies

And finally, in examining the results for mean radiant temperature, all sites show large reductions for $T_{mrt}$ under the scenario of additional trees. Mean radiant temperature is a useful proxy for thermal comfort as it essentially measures the mean of all surface temperatures in a given space and the thermal exchange with a particular point (e.g., a human body). The areas underneath trees are cooler by up to 19.5 °C at midday. While this is quite a significant result, it should also be noted that trees do have the potential for maintaining slightly higher $T_{mrt}$ than surrounding areas during evening and nighttime due to trapping of outgoing longwave radiation under the tree canopy. The models for this area show slightly elevated nighttime temperatures of 0.4–0.7 °C at 10 pm. This is a seemingly small trade-off when compared to the large daytime improvements in thermal comfort achieved with trees. During the day (at noon), $T_{mrt}$ can reach up to 72 °C while at 10 pm it is around 30 °C across the study area.

It is also worth noting that $T_{mrt}$ is increased or shows almost no improvement across all sites for the scenario of changing pavers and roads. This is interesting in that increasing the albedo (using light-colored surfaces) is often cited as a strategy for alleviating urban heat. However, while this certainly leads to improvement in surface temperatures, it may also reduce near-surface thermal comfort by increasing the $T_{mrt}$ near the ground surface (Erell et al. 2012). This result is supported by previous research showing that sheltered locations (either sheltered by vegetation or by surrounding buildings) show greatly reduced $T_{mrt}$ as compared to exposed locations (Ali-Toudert et al. 2005; Armson et al. 2012).

## 5.17   Designing Neighborhoods for Extreme Urban Heat

Considered together, the four scenarios presented in this research point toward several planning guidelines that can be adopted at neighborhood level for hot arid regions. Although the research may not be able to specifically determine how the physical changes are affecting Doha residents in terms of health and comfort, it describes the temperature patterns in three case study areas and provides guidance on alleviating urban heat, depending on thermal regimes in different areas of Doha.

One consideration is the temporal variation in temperatures associated with differing types of land cover. This may lead to opportunities for reducing temperatures during "shoulder periods," or transition times, during late morning and early evenings, which may offer chances for changing the urban design such that residents are able to spend more time outside. If planning agencies are considering options for mediating temperatures to provide pedestrians greater access to outdoor spaces, then reducing the amount of urban land cover (or impervious surfaces) may be the first step. While changing land cover may not be cost-effective or a feasible option in places containing a large amount of impervious surface, making space for trees within highly urbanized areas may be a reasonable alternative. Despite the arid climate, given the abundant amount of water from reusable sources such as air conditioning condensate (Bryant and Ahmed 2008) and treated sewage effluent (TSE) in Doha (Jasim et al. 2016), water resources for expanding the urban canopy may be readily available.

Although increasing the albedo of surfaces is a common practice, one that could reduce the absorption of solar radiation, it might not work for Doha and places with similar climatic conditions. Higher albedo could help the temperature in the morning and evening, but it could also have the adverse effect of increasing both the near-surface air temperature and mean radiant temperature during the daytime.

Finally, while the high-density high-rise development did estimate large surface temperature and $T_{mrt}$ reductions at midday, this only occurred in small areas directly in the shadow of buildings. It is important to consider that if coastal winds are blocked by high-rise buildings along the coastline, inland areas may not benefit from this mediating influence. Restricting development along the coast, especially those buildings that prevent these coastal processes from mediating inland temperatures, is a policy that has traction in scholarly research (Wong et al. 2011), and it may be a policy option that can improve the short- and long-term quality of life of Doha's residents. Also, the research indicates that lower density development may be the best solution for this climate, in that it allows for air circulation and shading between spaces through the use of natural, softscaped, communal spaces between buildings, and paved areas.

Again, at the neighborhood-scale, the most effective scenario was that of adding mature trees (approximately 10 m height, in 10 m intervals along the sides of roads). In the coastal site, the microclimate modeling estimated a maximum hourly air temperature reduction of 1.35 °C and surface temperature reductions in Site 5 up to 15 °C at 12 pm. While the scenario of changing the type of paver resulted in approximately 3 °C lower air temperatures compared to the base model for the urbanized site, we also found that comparing different areas of the base model showed that a standard light red paver will be lower by up to 12 °C than areas covered with asphalt and 3–4 °C cooler than areas of bare soil. However, $T_{mrt}$ was also increased or nearly neutral across all sites for the pavers and roads scenario. This result points to the need to develop improved shading measures for pedestrian pathways and outdoor recreational areas, especially for inland areas, as these experience the strongest UHI at midday when both air temperatures and $T_{mrt}$ are highest.

This study suggests a need for further and ongoing examination of Doha's urban heat and development patterns, particularly to gain additional insight on the specific planning measures that are most likely to lead to improvements in thermal comfort for residents. Such future research will continue to inform the planning choices, not only for Doha but the wider Middle East and North Africa region, and help to reduce the possible negative implications of development for humans and the environment upon which they depend.

## Appendix: Model Settings for Base Model Calibration

| Model version | Wind speed (10 m) | Initial temp | Spec hum | Rel hum | LBC | Soil wetness | Soil initial temp | Solar adjustment | Time steps | Switching angles | Timing (Plant, surface data, radiation) | Flow fields |
|---|---|---|---|---|---|---|---|---|---|---|---|---|
| Base V1 | 2 | 310 | 4 | 40 | Open | 30, 40, 40, 60 | 306, 304, 300, 293 | None | 20, 10, 5 | 30, 50 | 900, 60, 900 | 1800 |
| Base V2 | 4 | 310 | 4 | 40 | Open | 30, 40, 40, 60 | 306, 304, 300, 293 | None | 2, 2, 1 | 40, 50 | 600, 30, 600 | 900 |
| Base V3 | 5 | 308 | 6 | 45 | Open | 30, 40, 40, 60 | 306, 304, 300, 293 | 0.7 | 2, 2, 1 | 40, 50 | 600, 30, 600 | 900 |
| Base V4 | 4 | 307.39 | 4 | 45 | Simple Forcing | 30, 40, 40, 60 | 306, 304, 300, 293 | 0.9 | 2, 2, 1 | 40, 50 | 600, 30, 600 | 900 |
| Base V5 | 4 | 310 | 4 | 45 | Simple Forcing | 30, 40, 40, 61 | 306, 304, 300, 293 | 0.9 | 2, 2, 1 | 40, 50 | 600, 30, 600 | 900 |

# References

Ali-Toudert F, Djenane M, Bensalem R, Mayer H (2005) Outdoor thermal comfort in the old desert city of Beni-Isguen, Algeria. Clim Res 28(3):243–256. https://doi.org/10.3354/cr028243

Ali-Toudert F, Mayer H (2006) Numerical study on the effects of aspect ratio and orientation of an urban street canyon on outdoor thermal comfort in hot and dry climate. Build Environ 41(2):94–108. https://doi.org/10.1016/j.buildenv.2005.01.013

AlSarmi SH, Washington R (2014) Changes in climate extremes in the Arabian Peninsula: analysis of daily data. Int J Climatol 34(5):1329–1345. https://doi.org/10.1002/joc.3772

Armson D, Stringer P, Ennos ARR (2012) The effect of tree shade and grass on surface and globe temperatures in an urban area. Urban For Urban Green 11(3):245–255 (Elsevier GmbH). http://dx.doi.org/10.1016/j.ufug.2012.05.002

Arnfield AJ (2003) Two decades of urban climate research: a review of turbulence, exchanges of energy and water, and the urban heat island. Int J Climatol 23(1):1–26

Blazejczyk K, Epstein Y, Jendritzky G, Staiger H, Tinz B (2012) Comparison of UTCI to selected thermal indices. Int J Biometeorol 56(3):515–535 (Springer). https://doi.org/10.1007/s00484-011-0453-2

Bowler DE, Buyung-Ali L, Knight TM, Pullin AS (2010) Urban greening to cool towns and cities: A systematic review of the empirical evidence. Landsc Urban Plan 97(3):147–155 (Elsevier B.V.). https://doi.org/10.1016/j.landurbplan.2010.05.006

Bruse M (2011) ENVI-met model architecture. Mainz, Germany. http://www.envi-met.com/

Bruse M (2017) ENVI-met model architecture. http://envi-met.info/doku.php?id=intro:modelconcept#software_versions

Bruse M, Fleer H (1998) Simulating surface-plant-air interactions inside urban environments with a three dimensional numerical model. Environ Model Softw 13(3–4):373–384

Bryant JA, Ahmed T (2008) Condensate water collection for an institutional building in Doha, Qatar: an opportunity for water sustainability. Energy Systems Laboratory. http://esl.tamu.edu

Costello A, Abbas M, Allen A, Ball S, Bell S, Bellamy R, Friel S, Groce N, Johnson A, Kett M, Lee M, Levy C, Maslin M, Mccoy D, Mcguire B, Montgomery H, Napier D, Pagel C, Patel J, Puppim De Oliveira JA, Redclift N, Rees H, Rogger D, Scott J, Stephenson J, Twigg J, Wolff J, Patterson C (2009) Managing the health effects of climate change. Lancet 1693–1733

Das R (2017) Regulation of temperature of human body. https://www.slideshare.net/RanadhiDas1/regulation-of-temperature-of-human-body. Accessed 23 May 2018

Deng JY, Wong NH, Zheng X (2016) The study of the effects of building arrangement on micro-climate and energy demand of CBD in Nanjing, China. Procedia Eng 169:44–54. https://doi.org/10.1016/j.proeng.2016.10.006

El-Zein A, Jabbour S, Tekce B, Zurayk H, Nuwayhid I, Khawaja M, Tell T, Mooji YAl, De-Jong J, Yassin N, Hogan D (2014) Health and ecological sustainability in the Arab world: a matter of survival. Lancet 383(9915):458–476. https://doi.org/10.1016/S0140-6736(13)62338-7

Erell E, Pearlmutter D, Boneh D (2012) Effect of high-albedo materials on pedestrian thermal comfort in urban canyons. In: ICUC8—8th international conference on urban climates, Dublin, Ireland

ESRI (2017) Heatwaves: number of deadly heat days, ESRI maps: global risk of deadly heat. https://maps.esri.com/globalriskofdeadlyheat/#. Accessed 13 June 2018

Ferwati S, Skelhorn C, Shandas V, Voelkel J, Shawish A, Ghanim M (2018) Analysis of urban heat in a corridor environment—the case of Doha, Qatar. Urban Clim 24:692–702. https://doi.org/10.1016/j.uclim.2017.08.008

Gill SE (2006) Climate change and urban greenspace, School of Environment and Development, University of Manchester

Golden JS (2004) The built environment induced urban heat island effect in rapidly urbanizing arid regions—a sustainable urban engineering complexity. Environ Sci 1(4):321–349 (Taylor & Francis Group). https://doi.org/10.1080/15693430412331291698

Gunawardena KR, Wells MJ, Kershaw T (2017) Utilising green and bluespace to mitigate urban heat island intensity. Sci Total Environ 584–585:1040–1055 (The Author(s)). https://doi.org/10.1016/j.scitotenv.2017.01.158

Hathway EA, Sharples S (2012) The interaction of rivers and urban form in mitigating the urban heat island effect: a UK case study. Build Environ 58:14–22 (Elsevier Ltd.). https://doi.org/10.1016/j.buildenv.2012.06.013

Hong B, Lin B (2015) Numerical studies of the outdoor wind environment and thermal comfort at pedestrian level in housing blocks with different building layout patterns and trees arrangement. Renew Energy 73:18–27 (Pergamon). https://doi.org/10.1016/j.renene.2014.05.060

Jasim SY, Saththasivam J, Loganathan K, Ogunbiyi OO, Sarp S (2016) Reuse of treated sewage effluent (TSE) in Qatar. J Water Process Eng 11:174–182 (Elsevier). https://doi.org/10.1016/j.jwpe.2016.05.003

Kjelgren R, Montague T (1998) Urban tree transpiration over turf and asphalt surfaces. Atmos Environ 32(1):35–41 (Pergamon). https://doi.org/10.1016/s1352-2310(97)00177-5

Kyriakodis G-E, Santamouris M (2018) Using reflective pavements to mitigate urban heat island in warm climates—results from a large scale urban mitigation project. Urban Clim 24:326–339 (Elsevier). https://doi.org/10.1016/j.uclim.2017.02.002

Lelieveld J, Proestos Y, Hadjinicolaou P, Tanarhte M, Tyrlis E, Zittis G (2016) Strongly increasing heat extremes in the Middle East and North Africa (MENA) in the 21st century. Clim Change 23. https://doi.org/10.1007/s10584-016-1665-6

Leuzinger S, Vogt R, Körner C (2010) Tree surface temperature in an urban environment. Agric For Meteorol 150(1):56–62. https://doi.org/10.1016/j.agrformet.2009.08.006

Makido Y, Shandas V, Ferwati S, Sailor D (2016) Daytime variation of urban heat islands: the case study of Doha, Qatar. Climate 4(2):1–14

Matzarakis A, Rutz F, Mayer H (2007) Modelling radiation fluxes in simple and complex environments—application of the RayMan model. Int J Biometeorol 51(4):323–334 (Springer). https://doi.org/10.1007/s00484-006-0061-8

McPherson EG, Nowak DJ, Rowntree RA (1994) Chicago's urban forest ecosystem: results of the Chicago Urban Forest Climate Project. U.S. Department of Agriculture, Forest Service, Northeastern Forest Experiment Station, Radnor, PA, p 201

Oke TR (1976) The distinction between canopy and boundary layer urban heat islands. Atmosphere 14:268–277

Pal JS, Eltahir EAB (2016) Future temperature in southwest Asia projected to exceed a threshold for human adaptability. Nat Clim Change 6(2):197–200 (Nature Publishing Group). https://doi.org/10.1038/nclimate2833

Qatar General Secretariat for Development Planning (2011) Qatar National Development Strategy 2011–2016, Doha

Qatar Ministry of Municipality and Environment (2016) Qatar National Master Plan: Qatar National Development Framework 2032, Doha, Qatar

Rizzo A (2014) 'Rapid urban development and national master planning in Arab Gulf countries. Qatar as a case study. Cities 39:50–57. https://doi.org/10.1016/j.cities.2014.02.005

Santamouris M (2013) Using cool pavements as a mitigation strategy to fight urban heat island—a review of the actual developments. Renew Sustain Energy Rev. https://doi.org/10.1016/j.rser.2013.05.047

Santamouris M (2014) Cooling the cities—a review of reflective and green roof mitigation technologies to fight heat island and improve comfort in urban environments. Sol Energy 103:682–703 (Pergamon). https://doi.org/10.1016/j.solener.2012.07.003

Skelhorn C, Lindley S, Levermore G (2014) The impact of vegetation types on air and surface temperatures in a temperate city: a fine scale assessment in Manchester, UK. Landsc Urban Plann 121:129–140 (Elsevier B.V.). https://doi.org/10.1016/j.landurbplan.2013.09.012

Walikewitz N, Jänicke J, Langner M, Meier F, Endlicher W (2015) The difference between the mean radiant temperature and the air temperature within indoor environments: a case study during summer conditions. https://doi.org/10.1016/j.buildenv.2014.11.004

Watts N, Amann M, Ayeb-Karlsson S, Belesova K, Bouley T, Boykoff M, Byass P, Cai W, Campbell-Lendrum D, Chambers J, Cox PM, Daly M, Dasandi N, Davies M, Depledge M, Depoux A, Dominguez-Salas P, Drummond P, Ekins P, Flahault A, Frumkin H, Georgeson L, Ghanei M, Grace D, Graham H, Grojsman R, Haines A, Hamilton I, Hartinger S, Johnson A, Kelman I, Kiesewetter G, Kniveton D, Liang L, Lott M, Lowe R, Mace G, Odhiambo Sewe M, Maslin M, Mikhaylov S, Milner J, Latifi AM, Moradi-Lakeh M, Morrissey K, Murray K, Neville T, Nilsson M, Oreszczyn T, Owfi F, Pencheon D, Pye S, Rabbaniha M, Robinson E, Rocklöv J, Schütte S, Shumake-Guillemot J, Steinbach R, Tabatabaei M, Wheeler N, Wilkinson P, Gong P, Montgomery H, Costello A (2018) The Lancet Countdown on health and climate change: from 25 years of inaction to a global transformation for public health. Lancet 391(10120):581–630. https://doi.org/10.1016/S0140-6736(17)32464-9
Wong MS, Nichol J, Ng E (2011) A study of the "wall effect" caused by proliferation of high-rise buildings using GIS techniques. Landsc Urban Plan 102:(245–253)

# Chapter 6
# Predicting Urban Growth

**Yasuyo Makido, Vivek Shandas and Salim Ferwati**

**Abstract**  While scholars in North America, China, India, and Western Europe have examined the extent to which urban planning policies have contained growth, we still have yet to understand the role of urban growth containment strategies in mediating growth dynamics in non-Western regions of the world. We examine the case of Doha, Qatar to assess two future growth scenarios: (a) future growth, based on historical patterns of development; and (b) the extent to an urban growth boundary will mediate future growth. By using a combination of spatial analysis, and the application of a land transformation model, we address two research questions: (1) what landscape features help to explain the rapid expansion? And (2) how much land does an urban growth boundary help to conserve from future development? The results suggest that Doha has grown by approximately 213% between 1987 and 2013, and that if similar trends continue, the region will see another 26,350 hectares of urban development by 2028. We identify approaches for rapidly growing cities to apply modeling technique for evaluating the extent to which urban containment policies can support more sustainable forms of development.

**Keywords**  Rapid urbanization · Urban growth boundary · Land transformation model · Sustainability

One of the defining characteristics of the twenty-first century is the rapid and sustained growth of cities driven by foreign direct investments, megaprojects, and/or initiatives that aim to spur other forms of investment. As more of the world's population calls the city their home, the ability and capacity of planning agencies to balance the needs for growth with improvements in regional ecosystems, public investments, and human health, and well-being has become more complex and challenging. Since areas undergoing rapid urban growth often lack long-term and systematic information about the drivers and resulting patterns of growth, many planning agencies are forced to make decisions based on untested assumptions and incomplete or biased information (Cash et al. 2003).

V. Shandas et al., *Urban Adaptation to Climate Change*,
SpringerBriefs in Environmental Science,
https://doi.org/10.1007/978-3-030-26586-1_6

The physical form and growth of cities has been studied extensively over several decades and refers to the physical shape of a given area of development. Through the application of spatial analysis, we can, over a short period of time, evaluate the major trends of physical expansion of cities, and the extent to which specific policies (e.g., urban growth boundaries (UGBs), greenbelts, opens spaces, etc.) can shape future growth. Examples include the impact of development patterns from transportation corridors and innovation (Frenkel and Ashkenazi 2008), the process of growth and property values (Glaeser and Kahn 2004), density(Bertaud and Renaud 1997), and network configuration (Loutzenheiser 1997). The question of UGBs as an urban planning tool to contain growth is salient because many municipalities are implementing sustainability principles in their master plans. While UGBs have been used extensively to contain urban growth in the U.S., they have little application in rapidly developing regions of the world.

With their accompanying differences in land-use law, demographic shifts, and urban planning practices, land-use planners in non-Western countries have a lack of information for making critical decisions about growth. Tracing land-cover change can elucidate development patterns, characterize urban growth, and potentially reduce unintended social and environmental consequences, including excessive air pollution and oil consumption (Frumkin 2002; Glaeser and Kahn 2004; Brueckner and Helsley 2011), nutrient flows and biodiversity (Grimm et al. 2008), and high costs of building new public infrastructure and commercial developments (Hortas-Rico and Solé-Ollé 2010). Remotely sensed imagery has tracked the dramatic transformation of land surrounding cities from afar—Google's Earth Engine™ is one such example. Space-borne satellite data are particularly useful for developing countries due to the cost and time associated with traditional survey methods (Dong et al. 1997), and these techniques have become viable alternatives to conventional survey and ground-based urban mapping methods (Jensen et al. 2004; Dewan and Yamaguchi 2009).

Yet hidden from the camera are the complex local interaction of policy, management, economics, culture, and the surrounding environment that underlie the transformation of each new acre. Land-use change models (LUCM) provide some insight into these underlying drivers, which can then be used to project land transformation into the future or evaluate the influence of alternative policy and management scenarios on land-use and development patterns. Those cities that are undergoing rapid landscape transformation—which is the case for many Asian cities (Choi and Fricke 2010; Dewan and Yamaguchi 2009)—can be of particular interest to understanding the drivers, impacts, and mitigation options for containing urban growth because unlike their Western counterparts, which took several decades to create similar development patterns, Asian cities are transforming in a few years, if not months.

The control of growth is a fundamental function of local governments around the world. In many areas, such as the United States, these local policies are built on more than a century of case law. Many municipalities have found comprehensive plans as critical to integrating the various, disparate, and sometimes competing demands on local government. An increasing number of comprehensive plans include urban growth boundaries or greenbelts as ways to restrict the outward expansion of cities in favor of more costly infill. Unlike greenbelts, urban growth boundaries do not permanently restrict growth, but rather provide an interactive mechanism by which to coordinate infrastructure planning and growth demands.

For this study, we examined the spatiotemporal coupling of urban growth in one of the fastest urbanizing regions in the world: Doha, Qatar. Using data from satellite imagery, we analyzed physical changes in the Doha metropolitan region to provide a description of urban development patterns specific to that study region and extrapolate those patterns into the future. We address two research questions in the present study: (1) what landscape features help to explain the rapid expansion of Doha's metropolitan area? And (2) how much land does an urban growth boundary (UGB) affect future development patterns? We address these questions by examining two scenarios: (a) future growth, based on historical patterns of development; and (b) the extent to which future growth will extend into a potential UGB. By addressing these questions, we aim to contribute to the use of computer-based analysis of urban areas (e.g., urban geo-informatics) literature by describing the extent to which physical and spatial urban growths interact with temporal scales of development. We begin by offering a description of Doha and then outline our methods, including the use of remote-sensed images and the application of a LUCM approach to describe future urban growth trajectories. We conclude with a summary of our findings and identify some opportunities for future research.

## 6.1  Modeling Changes in Urban Land Cover

The placement and dimension of the urban growth boundary (UGB) was determined based on the Qatar National Development Framework (QNDF) 2032 strategic plan (Rizzo 2014). The QNDF specifically states a goal to, "Protect and enhance the natural environment, provide a long-term food security, and control urban sprawl to ensure that future urban development is contained within sustainable limits. This will be achieved through the introduction of metropolitan Doha greenbelt, where land has been allocated for the creation of a greenbelt around the urban area." Due to its rapid growth, lack of physical impediments, an abundance of cloud-free satellite data, Doha offers a timely and effective case study (Yin 2009) to evaluate the patterns and process of rapid growth, and the implications of imposing an urban growth boundary.

### 6.1.1  The Land Transformation Model

Of the available tools for characterizing drivers or urban growth, and predicting the implications of a UGB, arguably the most widely used approach is a land-use and cover change (LUCC) model. Land-use change models (LUCM) are used by researchers and professionals to explore the dynamics and drivers of these changes and to inform policies affecting such change. LUCM can be broadly classified into two groups: (a) "process-based" and (b) "empirically derived" models (Brown et al. 2012). Process-based models try to make explicit the multiple interactions between agents, organisms, and their environment. Because processes are explicit, these processes must be hardcoded. Agent-based models are an example of process-based models that has garnered particular attention of late. Empirically derived models estimate the temporal trends (in the case of longitudinal data) and/or spatial patterns (in the case of cross-sectional data) with a set of predictor variables. Such models do not depend on knowledge of social and natural processes driving the change, though are thought to influence land-use, including proximity to roads, proximity to cities and towns, mix of economic activity, demography, income and wealth, and biophysical factors like slope and soils.

Artificial neural networks (ANNs) are a type of empirically derived LUCMs based on machine learning techniques (Pijanowski et al. 2002). Neural networks are built from multiple layers of "perceptrons" connected by a series of weighted edges. The weighting of these edges is estimated iteratively using a set of training data to assess accuracy (Pijanowski et al. 2002, 2009). The training algorithm works by modifying weights until an evaluation metric (e.g., MSE) stabilizes. Like parametric regression techniques, neural networks can be used to predict outcome based on a set of known inputs. Unlike regression approaches, ANN does not assume that data vary according to a specific distribution (e.g., Gaussian) and are therefore able to fit highly complex and contingent patterns (Basheer and Hajmeer 2000).

The Land Transformation Model (LTM) is an empirical LUCM that relies on ANN to predict LUCC. The model has been used in numerous studies and in developing urban setting where development processes are not well understood (Pijanowski et al. 2002, 2009; Pithadia et al. 2005; Tayyebi et al. 2011; Tayyebi and Pijanowski 2014; Song et al. 2015). LTM combines a geospatial information system (GIS) with artificial neural networks (ANNs) to forecast LUCC based on a series of proxy variables (e.g., distance to roads). The ANN "learns" the relationship between past development patterns in the region underlying spatial variables (Pijanowski et al. 2002). Because cities are considering many options for continuing growth, yet few have evaluations about specific strategies, the use of LTM offers a timely and effective approach for informing potential growth scenarios.

### 6.1.2  Data and Analysis

Our analysis employs two land-use/land-cover (LULC) maps. LULC maps were derived from Landsat ETM satellite images which were acquired in 1987 and 2013. Land cover was classified into four categories: urban/built-up, vegetation, soil, and water (Shandas et al. 2017). These images were then reclassified to binary image, containing urban and nonurban classes (Fig. 6.1). In addition to these LULC maps, we also employed a digitized map with all the roads in the study region, which was provided by the Doha Ministry of Planning. The Landsat ETM images were the primary dataset for conducting the analysis, and the four categories indicate the

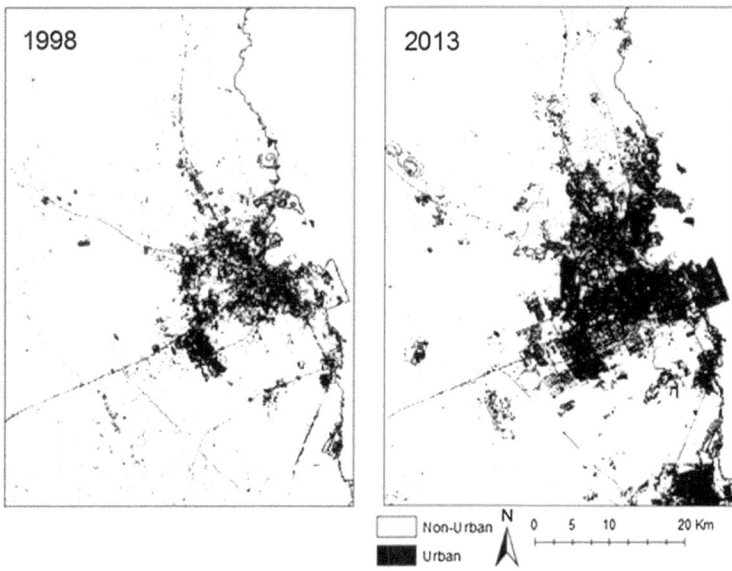

**Fig. 6.1** Urban area map in 1998 and 2013 (black represents urban, and white nonurban)

types of land cover at specific times of the study period, and how they change in subsequent years. Urban/built-up generally represents those areas that have asphalt concrete or another form of surface sealing; vegetation is generally everything from ground-level plants to large trees and everything in between; soil are those areas with cleared land though not containing other land covers, and water can be standing or moving.

Based on a plethora of existing studies that employ land transformation models, and our own prior research on land-cover change in the region, we employed five variables to simulate and predict LULC change: distance to roads (Cheng and Masser 2003; He et al. 2006; Batisani and Yarnal 2009; Shandas et al. 2017), distance to previously developed areas (Pijanowski et al. 2002; Hu and Lo 2007), distance to urban center (Cheng and Masser 2003), distance to vegetation (Tayyebi et al. 2011), and distance to the coast (Pijanowski et al. 2002). Although elevation and aspect have been used as important determinant, Doha is relatively flat and not impeded by impassable landforms or political boundaries. Using GIS (ArcMap 10.x), we resampled the five input layers to 90 × 90 m cell sizes (Fig. 6.2). The resulting layers result in a stacked raster consisting of a total 374,400 (=520 multiplied by 720) cells, which we used as our primary inputs for the LTM. The urban or "built-up" areas as of 1987 and water areas including ocean other water features were coded as exclusionary zones so that they would not undergo transitions. We used different input data to train and test the ANN to enable the LTM to have adequate predictive capacity. Training stage involves presenting input values and adjusting the

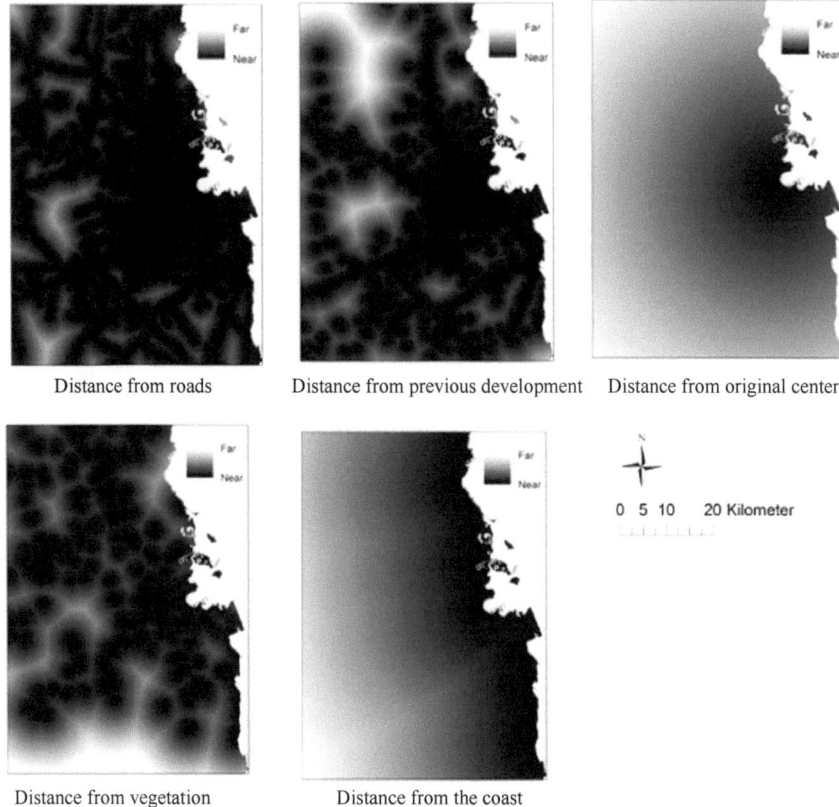

Distance from roads         Distance from previous development    Distance from original center

Distance from vegetation          Distance from the coast

**Fig. 6.2**  Maps of Doha in 1987 describing the five input variables

weights applied at each node according to the learning algorithm of backpropagation (Pijanowski et al. 2002, 2014; Song et al. 2015). The backpropagation algorithm randomly selects the initial weights and then compares the calculated output for a given observation with the expected outcome for that observation (Pijanowski et al. 2002). In this project, we trained ANNs using the actual urban changes between 1998 and 2013 (Fig. 6.2).

## 6.1.3  Simulating Urban Expansion

We developed two scenarios for evaluating the implications of expanding or constraining urban growth. The first scenario included the original input variables, and subsequent urban growth unimpeded through 2028. The second growth scenario consisted of urban growth although we imposed an urban growth boundary (Fig. 6.3). The UGB was digitized using the Qatar National Development Framework, which

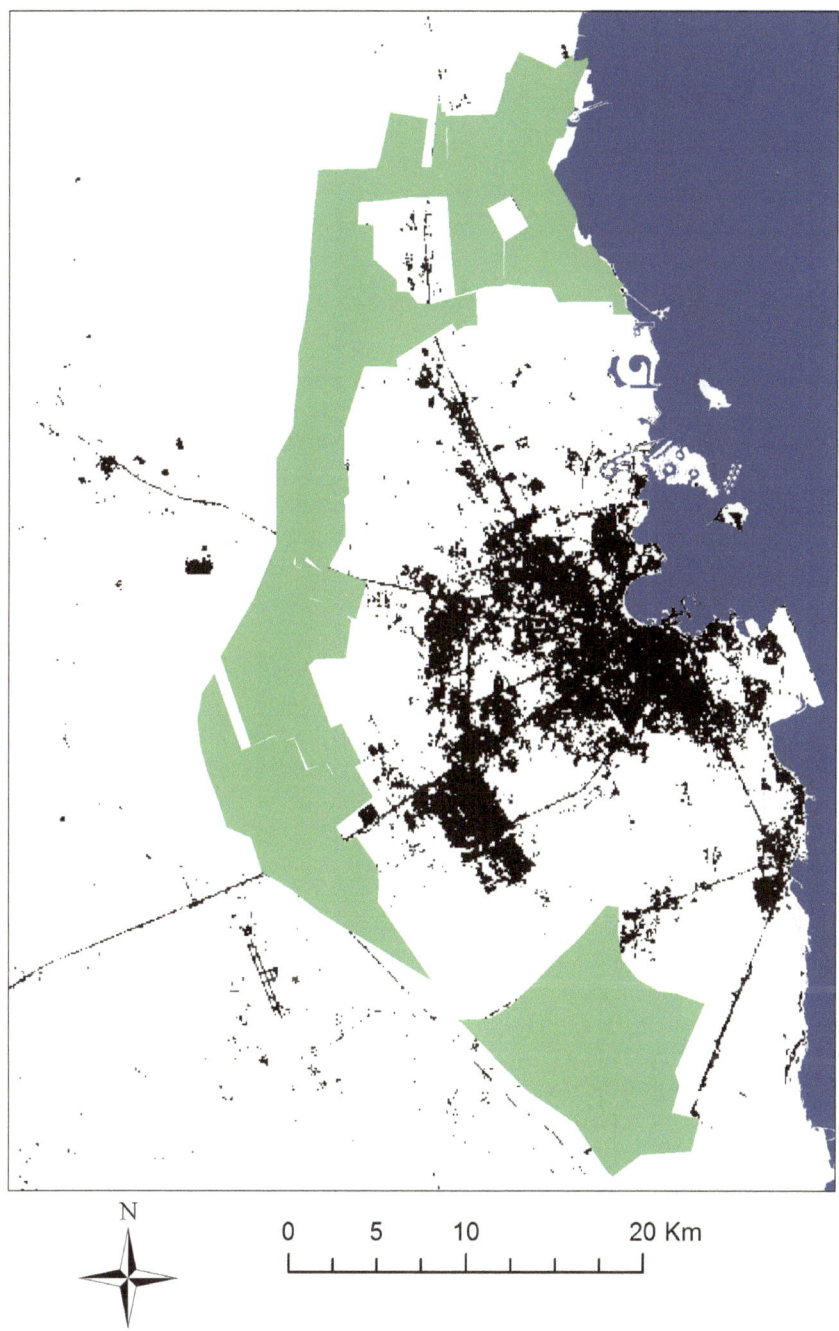

**Fig. 6.3** Urban area in 1998 and the greenbelt

contained a specific location for the "greenbelt." Using existing data from earlier analysis (Shandas et al. 2017), we calculated the rate, location, and quality of urban growth in the Doha metropolitan region from 1987 through 2013. To forecast to 2028, we assumed that the expansion rate of the urban area between 2013 and 2028 will be same as between 1998 and 2013 and that geographic spatial drivers that have influenced development patterns in the past will remain similar in the future.

### 6.1.4   Evaluating Model Performance

Using historical growth patterns to predict future growth is a well-accepted approach in the literature of planning and geography. Evaluating model performance can improve the likelihood of accurately forecasting future growth. We evaluated model performance using two methods: Percent Correct Metric (PCM) and Relative Operating Characteristics (ROC). PCM is the ratio between the number of cells having changes that are correctly predicted and the total number of cells that have undergone changes during the research period (Pijanowski et al. 2002; Pijanowski et al. 2006; Song et al. 2015).

$$\text{PCM} = \frac{(\text{\# of cells correctly predicted to change}) \times 100.0}{(\text{\# of cells that actually transitioned})}$$

ROC also known as receiver operating characteristic is used with area under curve (AUC). ROC is a statistical method to measure the association between presence and absence of characteristic (Egan 1975; Fawcett 2006) and has been used to measure relationship between simulated change and real change in land-cover change modeling (Pontius and Schneider 2001; Pontius and Parmentier 2014; Pontius and Si 2014). The area under curve (AUC) is the summary metric of the ROC, and it can range from 0.5 to 1, with higher values indicating greater levels of model prediction, and values near 0.5 indicate the random allocation. We used TOC Web App (TOC no date), which is the R Shiny web application to construct the ROC curve and compute AUC value for this study.

We also examined the importance of individual predictor variables by iteratively removing each variable and evaluating changes in the mean squared error (MSE). The MSE is the difference between the expected and calculated output values across all observations (Tayyebi et al. 2011). Variables with larger MSE values explained a greater degree of past development patterns (Fig. 6.4).

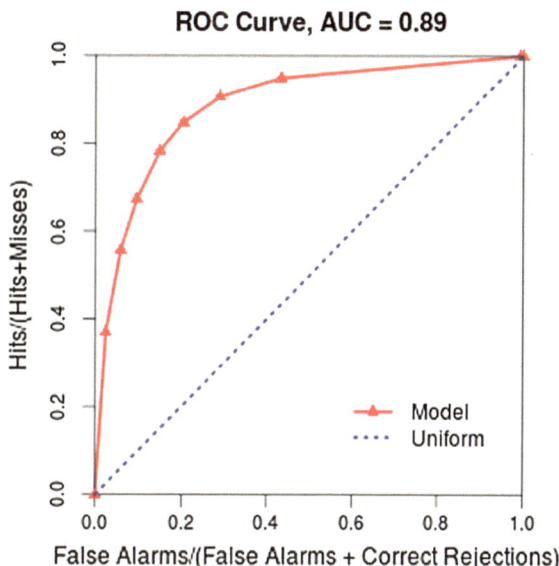

**Fig. 6.4** ROC curve for Scenario 2

## 6.2 Assessing Change in Land Cover

To evaluate the effectiveness of our growth model, we created four categories of urban expansion during the periods of 1998 to 2013 in the study area (Fig. 6.5). The four categories indicate no real change, no predicted change (True Negative); no real change, but change predicted by the model (False Negative); real change and predicted change (True Positive); and real change but not predicted by the model (False Positive). When evaluating these results, we note that the model failed to simulate urban areas (False Positive) when the areas were located relatively far from the original center.

### 6.2.1 Land Transformation Model Performance

In the training run of the LTM for Scenario 1 (S1), the MSE started at 0.0597 and after 3000 cycles, decreased to values around 0.0581 after 3000 cycles. As previous studies added more cycles after the model was stabilized (Pijanowski et al. 2002; Song et al. 2015), we halted the training at 6000 cycles, when the MSE was 0.0581 for S1. The same number of cycles was applied to Scenario 2 (S2). The number of cells correctly predicted to change and the number of cells that actually changed were, respectively, 27,387 and 38,512 for S1, and 25,253 and 32,530 for S2. The resulting PCM of the model was 71.11% for S1 and 77.63% for S2. As a general rule of thumb, models with PCM between 40 and 60% are considered acceptable, while models with PCM between 60 and 80% are considered to be an exceptional

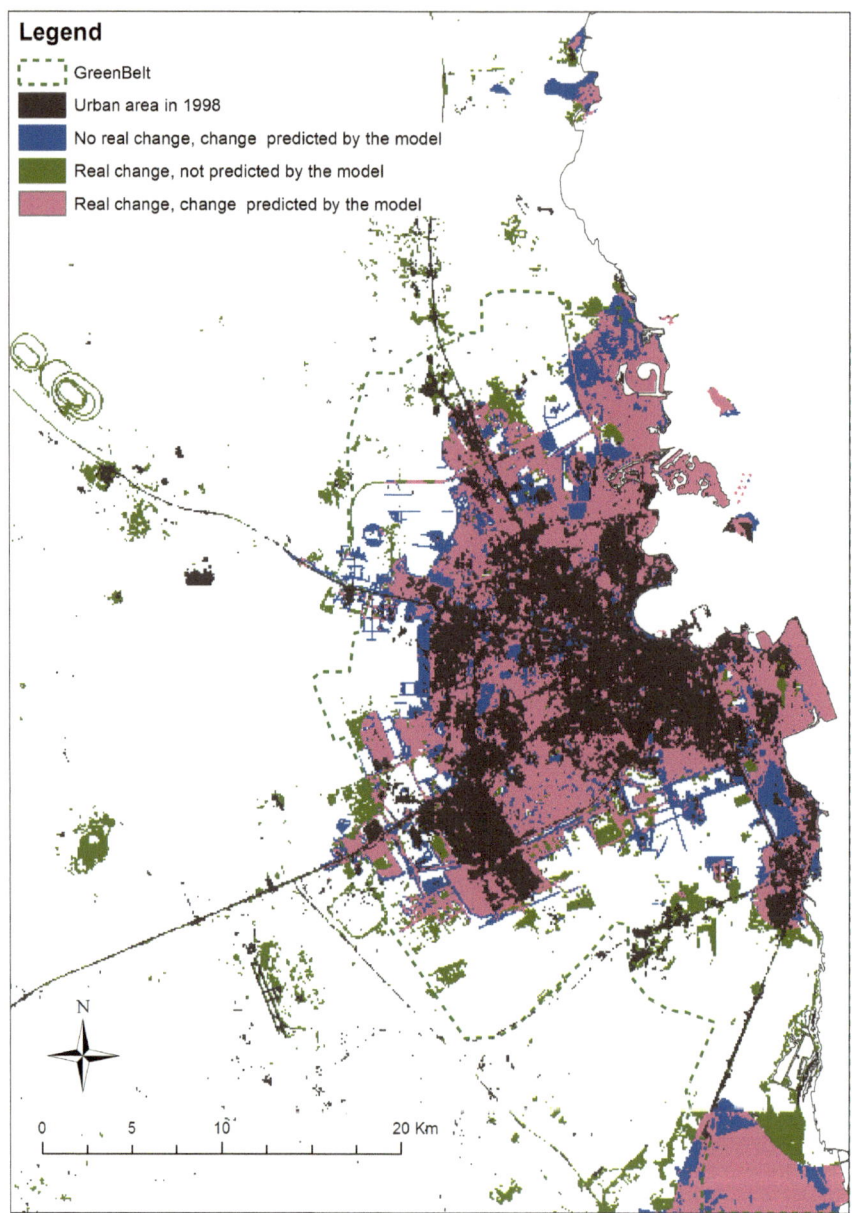

**Fig. 6.5**  Simulation (S1) results of urban expansion during 1998–2013 in Doha

**Table 6.1** Predictor variables, MSE values, and level of importance in predicting future urban growth in the Doha metropolitan region

| Reduced variable | MSE value | Difference to MSE of using five variables (0.00581) | Importance |
|---|---|---|---|
| Distance from urban center | 0.0737 | 0.0156 | 1 |
| Distance from the coast | 0.0726 | 0.0144 | 2 |
| Distance from urban areas | 0.0644 | 0.0063 | 3 |
| Distance from vegetation | 0.0641 | 0.0060 | 4 |
| Distance from roads | 0.0629 | 0.0048 | 5 |

(Pijanowski et al. 2006). The AUC for S1 was 0.92 and S2 was 0.89 (Fig. 6.5). Although we know of no definite values to evaluate model performance, the AUC ranges from 0.5 (random) to 1.0 (perfect), to which we consider our results to be acceptable.

## 6.2.2 Relative Effect of Predictor Variables

The MSEs of the five input variables describe the importance of each (Table 6.1). The difference in MSE value indicates the level of impact that each additional layer has on the predictability of the model. Based on the MSE, the least important variable was distance to road, while the most important variable was distance to original center, which we defined as the historic Doha center (Khalil and Shaaban 2012). While all these factors were important in predicting growth, the MSE values suggest that outward growth from Doha's original area was the most important in determining those locations where it grew in the future. While road, city centers, and coastal areas have been shown to be predictors of growth in other models (Pijanowski et al. 2006; Tayyebi and Pijanowski 2014); unexpectedly, distance to vegetation also emerged as one of the top five contributors to growth. Vegetation in this case may suggest a location with sufficient plumbing and water infrastructure, which, we speculate, makes it an attractor to development.

## 6.2.3 Forecasting Urban Growth

In examining future growth scenarios, we examined the aforementioned four categories. The results indicate that the model failed to simulate urban areas (False Positive) when the areas were located relatively far from the original center (Fig. 6.5).

Between 1998 and 2013, the simulated urban area of Doha increased by 285 km$^2$ within S1 and 263 km$^2$ within S2 (Table 6.2). Approximately, 7% (3,450 hectares) of the growth between 1980 and 2013 in Doha occurred outside of the UGB, the major-

**Table 6.2** Predictor variables, MSE values, and level of importance in predicting future urban growth in the Doha metropolitan region

| Measure of the urban area (hectares) | Scenario 1 (S1) | Scenario 2 (S2) (within boundary) |
|---|---|---|
| Total urban area in 1998 | 21,632 | 20,401 |
| Total urban area in 2013 | 50,204 | 46,751 |
| Forecasted total urban area in 2028 | 84,021 | 73,100 |
| Vegetation converted to urban in (2013–2028) | 786 | 674 |
| Soil converted to urban (2013–2028) | 34,309 | 26,344 |
| Total nonurban converted to urban (2013–28) | 35,095 | 27,018 |

ity of which occurred along major transportation corridors or highways. In general, the growth seems to occur as a result of infrastructure investment on highways in particular. The scenario S1 overpredicts growth and underpredicts highway and leapfrog development. Toward 2028, the total amount of land conserved is 10,920 hectares or almost 13% of the landscape. These results also suggest that a UGB may reduce the amount of vegetation by 2028—most of which are constructed and not naturally occurring—and could be replaced in locations where surfaces do not preclude planting spaces.

A major difference between the two scenarios is that S2 has more infill development than S1, especially in the southeast and northeast of the metropolitan region (Fig. 6.6). These areas are particularly prone to development, in part because of the important role of coastal development over the past 25 years. Infill areas also preclude growth because the model would force growth to occur in previously undeveloped areas.

## 6.3  Implication for Urban Growth Boundaries

The role of urban growth boundaries in containing rapid development is a largely unexplored area of research. While such "smart growth" policies are present in many parts of the world, few have tested their effectiveness. Often planning decisions are made iteratively based on immediate knowledge, policies, and political pressures. Without systematic assessment of the role of specific policies in mediating growth patterns across a region, rapidly urbanizing cities can create unsustainable living environments with costly impacts on human health and local ecologies. While formal urban growth containment policies have in use since the 1970s, few studies have examined the extent to which they contain growth. In rapidly urbanizing regions, we can find opportunities to apply recent advances in geospatial techniques and modeling for weighing alternative development scenarios.

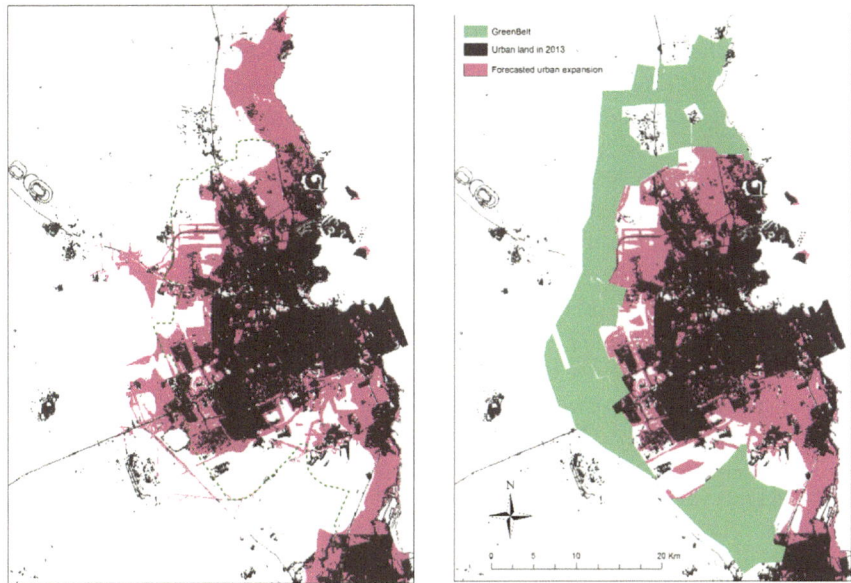

**Fig. 6.6**  Urban area in 2028 forecasted by LTM (Left: S1, Right: S2)

Assuming that such a boundary would be effectively implemented, our findings suggest that extensive areas of land can be conserved over a short period of time. Although the most important drivers of growth are the distance to city center and coast, respectively, we also observe that as areas develop, they create new roads and urban areas, both of which can amplify future expansion. The distance to vegetation serves as an important contributor to growth of Doha and may suggest that, as areas develop, they bring with them additional vegetation, which is often scarce in a desert landscape. In addition, since the distance from the city center is the most important predictor when using past dates (1987–2013), and urban growth occurs at greater distances from the city center, we expected the model to underperform in those distant areas of the metro region. If the UGB was instituted in the earlier part of that period, the model suggests a reduction of 3,475 hectares of land developed.

The differences between 2013 and 2028 offer many insights into the potential effectiveness of a UGB in conserving landscapes from urban development. First, the UGB conserved almost 11 thousand hectares of land, which in and of itself suggests a major role for urban containment systems, if effectively implemented. Second, the model also indicates that the total vegetation would be reduced, perhaps due to the extensive infill development occurring between 2013 and 2028 (S2). While the model assumes current development patterns, the presence of vegetation often relies on the design of buildings, and those that allow for open space can maintain or increase vegetation amounts. Third, the model outputs are spatially explicit and describe conservation of coastlines, which account for the majority (57%) of the conservation areas from the UGB.

Perhaps equally important, although outside the scope of the present project, the conservation of large areas of land can also create major saving to infrastructure costs, reduce pollution from commuting, and improve health benefits from limiting isolation and increasing walkability. Extensive evidence over the past 40 years indicate that smart growth policies, of which UGBs are one instrument, reduce overall costs of infrastructure between 38 and 50% (Fulton et al. 2013). Most of these expenditures go to developing new school buildings, roads and highways, water and sewer facilities, libraries, and utilities, while a smaller portion goes to other provides such as public services (e.g., police, fire, utility services, etc.). As fiscal capacity contacts with increases in oil prices and other provisions, these savings will likely mean greater resources can be devoted to other services. Moreover, the creation of greater density can also increase economic activity within specific areas where communities have greater access to local services (Choi and Fricke 2010).

Analytically, most studies use a contingency table in two categories, change and no change for assessing the model for binary classification (Tayyebi and Pijanowski 2014). In this study, we calculated PCM based on the contingency table for assessing the model performance, which helps us to validate findings. In addition, we examined a statistical method, ROC, to measure the association between presence and absence of characteristic. The results of both PCM and ROC confirm the adequacy of modeling. At the same time, although no model will be completely accurate, we view these results as a means for examining what alternative development strategies may improve sustainability outcomes, one begins the fiscal health of the region.

Another such analytical benefit of the present study is a case study to explore modeling applications to planning practice, in this case for the evaluation of the effectiveness of specific urban containment policies. Such modeling exercises are commonplace in municipal planning organizations (MPOs), though currently focus on land-use and transportation modeling, hydrology, and/or environmental change. Besides their use in anticipating growth demand and scheduling installation and maintenance of infrastructure, urbanization models can also build institutional capacity for controlling growth by building consensus over future planning and goals and tightening coordination between different agencies and departments. This is particularly critical in rapidly growing cities that lack institutional capacity to address growth, particularly in the face of huge capital investments. Anticipating future development is critical because unlike other forms of land-use change, urbanization is often permanent and can entail massive differed costs. The rapid changes in demographics and infrastructure cause spillover impacts on neighboring land, a cycle which can propagate outward from the city core with surprising rapidity.

Our assessment also contains limitations. One such limitation is that we have created a technocratic view that describes the process for growth. The increased availability of remotely sensed data provides key insights into this process, though essential to these growth processes is the fact that the result occurs due to a myriad of different influences that create the city. The various forces that drive urbanization are global and couched in local polices that mediate the likelihood of development. We assume in future years that these policies will remain the same, which is unlikely. At the same time, modeling urban growth using LUCM reveals something about the context where urban growth has occurred, and by extension, insight into future growth possibilities. Such predictions only represent possibilities and must only be considered with the assumptions embedded in the modeling exercise.

## 6.4   Urban Growth Scenarios for Doha

In this study, we evaluated two scenarios of urban growth in one rapidly urbanizing region of the world. Among the most rapidly growing areas in the world, Doha offers a timely and meaningful approach to evaluating one approach, an urban growth boundary. Even with relatively simple articulations of urban growth patterns—based on plausible scenarios—decision-makers and others can visualize and quantify the implications of an urban growth boundary. While a general lack of relevant data still hinders comprehensive assessments of development patterns in the Middle East, recent developments in spatial analysis, computational power, and publicly available satellite data offer a timely and effective means for understanding growth of Middle East cities. For example, assessing urban growth over 50 years in the U.S. can conflate the drivers of growth, such as the rationale for the growth of specific areas and the outcomes of alternative growth patterns. When examining physical growth patterns over a shorter period of time during a phase of rapid urban development, we can, arguably, link specific growth processes and reasons for specific patterns of development. We can address such questions as: How does physical growth couple with existing infrastructures to create specific development patterns? How might new forms of development mitigate or exacerbate the environmental and human health impacts of urban growth? The rapid growth occurring in Middle East cities provides an opportunity to answer some of these questions by capturing data over short periods of fast development.

Each new acre transformed into urban use is the result of the complex local interaction of policy, management, economics, culture, and the surrounding environment—processes hidden from the camera. Land-use change models (LUCM) provide some insight into these underlying drivers, which can then be used to project land transformation into the future or evaluate the influence of alternative policy and management scenarios on land-use and development patterns. These topics are central to challenges facing the sustainability of urban regions, especially as population growth and climate change amplify already vulnerability infrastructure and ecologies. The novelty of urban areas and their sustainability requires at once a recognition of the features on the landscape that impact physical and social process and information for guiding the growth.

These environmental and social problems, which are often exacerbated by the physical patterns of urban development, point to the need for a better understanding of how urban areas outside of American and European contexts grow. In the past few years, researches in China, India, and Latin America have provided insight into non-Western development patterns (Monkkonen 2008; Li et al. 2009; Bhatta et al. 2010), but these studies are largely untested in the Middle East, which early evidence indicates contain a combination of Western and Eastern development models (Wiedmann 2012; Nassar et al. 2014; Rizzo 2014). As we better understand the processes of urban growth, we can more likely anticipate potential impacts from specific forms of development. These relationships between urban patterns of growth and impact on social and environmental problems offer a means for coupling land-use modeling exercises such as the present study with "dose–response" systems that allow decision-makers to better address current sustainability challenges.

# References

Basheer IA, Hajmeer M (2000) Artificial neural networks: fundamentals, computing, design, and application. J Microbiol Methods (Neural Comput Microbiol) 43(1):3–31. https://doi.org/10.1016/S0167-7012(00)00201-3

Batisani N, Yarnal B (2009) Urban expansion in Centre County, Pennsylvania: spatial dynamics and landscape transformations. Appl Geogr 29(2):235–249. https://doi.org/10.1016/j.apgeog.2008.08.007

Bertaud A, Renaud B (1997) Socialist cities without land markets. J Urban Econom 41(1):137–151. https://doi.org/10.1006/juec.1996.1097

Bhatta B, Saraswati S, Bandyopadhyay D (2010) Urban sprawl measurement from remote sensing data. Appl Geogr (Climate Chang Appl Geogr Place Policy Practice) 30(4):731–740. https://doi.org/10.1016/j.apgeog.2010.02.002

Brown DG, Walker R, Manson S, Seto K (2012) Modeling land use and land cover change. Land change science. Springer, Dordrecht (Remote Sensing and Digital Image Processing), pp 395–409

Brueckner JK, Helsley RW (2011) Sprawl and blight. J Urban Econom 69(2):205–213. https://doi.org/10.1016/j.jue.2010.09.003

Cash DW, Clark WC, Alcock F, Dickson NM, Eckley N, Guston DH, Jäger J, Mitchell RB (2003) Knowledge systems for sustainable development. Proc Natl Acad Sci 100(14):8086–8091. https://doi.org/10.1073/pnas.1231332100

Cheng J, Masser I (2003) Urban growth pattern modeling: a case study of Wuhan city, PR China. Landscape Urban Plann 62(4):199–217. https://doi.org/10.1016/S0169-2046(02)00150-0

Choi K, Fricke C (2010) Analyzing the impact of smart growth on projected road development in 2030

Dewan AM, Yamaguchi Y (2009) Using remote sensing and GIS to detect and monitor land use and land cover change in Dhaka Metropolitan of Bangladesh during 1960–2005. Environ Monit Assess 150(1–4):237. https://doi.org/10.1007/s10661-008-0226-5

Dong Y, Forster B, Ticehurst C (1997) Radar backscatter analysis for urban environments. Int J Remote Sens 18(6):1351–1364. https://doi.org/10.1080/014311697218467

Egan J (1975) Signal detection theory and ROC analysis. Academic Press

Fawcett T (2006) An introduction to ROC analysis. Pattern Recogn Lett (ROC Anal Pattern Recogn) 27(8):861–874. https://doi.org/10.1016/j.patrec.2005.10.010

Frenkel A, Ashkenazi M (2008) Measuring urban sprawl: how can we deal with it? Environ Plan 35(1):56–79. https://doi.org/10.1068/b32155

Frumkin H (2002) Urban sprawl and public health. Public Health Rep 117(3):201–217. https://doi.org/10.1093/phr/117.3.201

Fulton W, Preuss I, Dodds A, Absetz S, Hirsch Fulton P (2013) Building better budgets: a national examination of fiscal benefits to smart growth development. Smart Growth of America

Glaeser EL, Kahn ME (2004) Chapter 56—sprawl and urban growth. In: Henderson JV, Thisse J-F (eds) Handbook of regional and urban economics. Elsevier (Cities and Geography), pp 2481–2527

Grimm NB, Foster D, Groffman P, Grove JM, Hopkinson CS, Nadelhoffer KJ, Pataki DE, Peters DP (2008) The changing landscape: ecosystem responses to urbanization and pollution across climatic and societal gradients. Front Ecol Environ 6(5):264–272. https://doi.org/10.1890/070147

He C, Okada N, Zhang Q, Shi P, Zhang J (2006) Modeling urban expansion scenarios by coupling cellular automata model and system dynamic model in Beijing, China. Appl Geogr 26(3–4):323–345. https://doi.org/10.1016/j.apgeog.2006.09.006

Hortas-Rico M, Solé-Ollé A (2010) Does urban sprawl increase the costs of providing local public services? evidence from Spanish municipalities. Urban Stud 47(7):1513–1540. https://doi.org/10.1177/0042098009353620

Hu Z, Lo CP (2007) Modeling urban growth in Atlanta using logistic regression. Comput Environ Urban Syst 31(6):667–688. https://doi.org/10.1016/j.compenvurbsys.2006.11.001

Jensen RR, Gatrell JD, Boulton JR, Harper BT (2004) Using remote sensing and geographic information systems to study urban quality of life and urban forest amenities. Ecol Soc 9(5):5

Khalil RF, Shaaban K (2012) Rebuilding old downtowns: the case of Doha, Qatar. REAL CORP 2012 proceedings, 0(May 2012), pp 677–689. https://doi.org/10.13140/2.1.3296.1600

Li F, Liu X, Hu D, Wang R, Yang W, Li D, Zhao D (2009) Measurement indicators and an evaluation approach for assessing urban sustainable development: a case study for China's Jining City. Landscape Urban Plann 90(3):134–142. https://doi.org/10.1016/j.landurbplan.2008.10.022

Loutzenheiser D (1997) Pedestrian access to transit: model of walk trips and their design and urban form determinants around bay area rapid transit stations. Transport Res Record J Transport Res Board 1604:40–49. https://doi.org/10.3141/1604-06

Monkkonen P (2008) Using online satellite imagery as a research tool: mapping changing patterns of urbanization in Mexico. J Plann Educat Res 28(2):225–236. https://doi.org/10.1177/0739456X08323771

Nassar AK, Alan Blackburn G, Duncan Whyatt J (2014) Developing the desert: the pace and process of urban growth in Dubai. Comput Environ Urban Syst 45:50–62. https://doi.org/10.1016/j.compenvurbsys.2014.02.005

Pijanowski BC, Alexandridis KT, Müller D (2006) Modelling urbanization patterns in two diverse regions of the world. J Land Use Sci 1(2–4):83–108. https://doi.org/10.1080/17474230601058310

Pijanowski BC, Brown DG, Shellito BA, Manik GA (2002) Using neural networks and GIS to forecast land use changes: a land transformation model. Comput Environ Urban Syst 26(6):553–575. https://doi.org/10.1016/S0198-9715(01)00015-1

Pijanowski BC, Tayyebi A, Delavar MR, Yazdanpanah MJ (2009) Urban expansion simulation using geospatial information system and artificial neural networks. Int J Environ Res 3(4):493–502

Pijanowski BC, Tayyebi A, Doucette J, Pekin BK, Braun D, Plourde J (2014) A big data urban growth simulation at a national scale: configuring the GIS and neural network based Land Transformation Model to run in a High Performance Computing (HPC) environment. Environ Model Softw 51:250–268. https://doi.org/10.1016/j.envsoft.2013.09.015

Pithadia S, Shellito BA, Alexandridis K (2005) Calibrating a neural network-based urban change model for two metropolitan areas of the Upper Midwest of the United States. Int J Geogr Inf Sci 19(2):197–215. https://doi.org/10.1080/13658810410001713416

Pontius RG, Parmentier B (2014) Recommendations for using the relative operating characteristic (ROC). Landscape Ecol 29(3):367–382. https://doi.org/10.1007/s10980-013-9984-8

Pontius RG, Schneider LC (2001) Land-cover change model validation by an ROC method for the Ipswich watershed, Massachusetts, USA. Agric Ecosyst Environ. (Predict Land-Use Change) 85(1):239–248. https://doi.org/10.1016/s0167-8809(01)00187-6

Pontius RG, Si K (2014) The total operating characteristic to measure diagnostic ability for multiple thresholds. Int J Geogr Inf Sci 28(3):570–583. https://doi.org/10.1080/13658816.2013.862623

Rizzo A (2014) Rapid urban development and national master planning in Arab Gulf countries. Qatar as a case study. Cities 39:50–57. https://doi.org/10.1016/j.cities.2014.02.005

Shandas V, Makido Y, Ferwati S (2017) Rapid urban growth and land use patterns in Doha, Qatar: opportunities for sustainability? Eur J Sustain Develop Res 1(2):1–13. https://doi.org/10.20897/ejosdr.201711

Song W, Pijanowski BC, Tayyebi A (2015) Urban expansion and its consumption of high-quality farmland in Beijing, China. Ecol Indicat 54:60–70. https://doi.org/10.1016/j.ecolind.2015.02.015

Tayyebi A, Pijanowski BC (2014) Modeling multiple land use changes using ANN, CART and MARS: Comparing tradeoffs in goodness of fit and explanatory power of data mining tools. Int J Appl Earth Obs Geoinf 28:102–116. https://doi.org/10.1016/j.jag.2013.11.008

Tayyebi A, Pijanowski BC, Tayyebi AH (2011) An urban growth boundary model using neural networks, GIS and radial parameterization: an application to Tehran, Iran. Landscape Urban Plann 100(1–2):35–44. https://doi.org/10.1016/j.landurbplan.2010.10.007

TOC (no date)

Wiedmann F (2012) Post-oil urbanism in the Gulf: new evolutions in governance and the impact on urban morphologies

Yin RK (2009) How to do better case studies. In: The SAGE handbook of applied social research methods, vol 2. SAGE, pp 254–282

# Chapter 7
# Conclusion: Livability in a Warming Climate

Vivek Shandas

**Abstract** Taken together, the studies presented for Qatar point toward a number of planning guidelines that can be adopted at different scales, ranging from regional down to neighborhood level. While the recent physical development of Qatar is discernable through an analysis of satellite images, the quality of growth and its implications for the social, economic, and environmental conditions of the region all require further study and analysis. These studies have analyzed the patterns of urbanization that led to significant challenges in thermal comfort and provide guidance on alleviating urban heat, depending on thermal regimes in different areas of Doha. Findings suggest that the the built enviornment can, in fact, mediate extreme urban heat, though likely only at the 'shoulder periods', which consist of periods in the morning and evening. The ability for creating outdoor spaces that are inviting and habitable is essential for livability. Though these chapters are an early examination of the opportunities for improving urban livability, they are, admittedly, an early exploration. Future work will need to find a way to refine further the models, analyses, and ultimate applications of this work. Without doubt, the field of urban climate adapation sciences will provide a springboard for society to advance livability on a burning planet.

**Keywords** Livability · Doha · Outdoor · Built environment

While increased climate change warnings seem to be proclaimed on a far more regular basis, cities worldwide are growing at faster rates than ever before. Rapid population growth brings greater urban density which means more buildings made of materials that in turn generate and radiate more heat resulting in increased Urban Heat Island effect. And these indisputable facts compound as global temperatures rise. Yet, the capacity of humans and the built environment to withstand increases in extreme temperatures is still largely unknown. We know that at 37 °C, the human body registers thermal discomfort and 38 °C marks the threshold to physiological distress. Though unlike humans, who can potentially increase tolerance to spending more of our lives confronting higher temperatures, the extent to which the complex set

V. Shandas et al., *Urban Adaptation to Climate Change*,
SpringerBriefs in Environmental Science,
https://doi.org/10.1007/978-3-030-26586-1_7

of mechanical systems, gray infrastructure, and network of pipes and other conduits are able to handle the changes in temperature will require careful study. Ultimately, indicators of human health will describe the extent to which our built environment is serving society. There are an increasing number of cities around the planet whose temperature and humidity ranges challenge the ability of residents to find inviting levels of comfort out of doors during hot periods of the year. Doha, Qatar is among them, and its residents are not alone in living with these conditions. As the number of these places grows, urban adaptation researchers increasingly believe that places that are enduring the hottest of temperatures can help us learn the strategies that can improve urban sustainability. Doha is a case study that few other cities can rival, at least from the perspective of living with heat, rapid urbanization, and exploring prospects for livability. This book provides early observations, analyses, and findings about, and recommendations for the enhancement of urban places and spaces in the public realm. Our premise for writing this work is that outdoor enjoyment of safe and inviting public places by all should be a defining condition of urban livability, and that their provision and maintenance should be considered a communal responsibility of national, municipal, and local governments for which users should be expected to contribute in fact and in kind to their programming and use.

Taken together, the studies presented for Qatar point toward a number of planning guidelines that can be adopted at different scales, ranging from regional down to neighborhood level. While the recent physical development of Qatar is discernable through an analysis of satellite images, the quality of growth and its implications for the social, economic, and environmental conditions of the region all require further study and analysis. These studies have analyzed the patterns of urbanization that led to significant challenges in thermal comfort and provide guidance on alleviating urban heat, depending on thermal regimes in different areas of Doha.

One consideration is the temporal variation in temperatures associated with differing types of land cover. This may lead to opportunities for reducing temperatures during "shoulder periods" or transition times during the late morning and early evenings, which may offer chances for changing the urban design such that people can spend more time outside. This chapter reviews all the evidence gathered through different studies presented and develops guidance and strategies for planning agencies to consider in mediating temperatures to provide pedestrians greater access to outdoor spaces and greater choice in modes of transport. Despite the periods of extreme heat and arid climate, strategic planning and management of urban areas can improve residents' and visitors' ability to live, work, and move throughout the city comfortably.

The four studies presented in this book point toward a number of planning guidelines that can be adopted at different scales, ranging from regional down to neighborhood level. While the physical growth of Doha is discernable through an analysis of satellite images, the quality of growth and its implications for the social, economic, and environmental conditions of the region all require further study. Although the studies may not be able to specifically determine how the physical changes are affecting Doha residents, they can describe the patterns of urbanization that led to significant challenges in other regions of the world and provide some guidance on alleviating urban heat, depending on thermal regimes in different areas of Doha.

One consideration is the temporal variation in temperatures associated with differing types of land cover. This may lead to opportunities for reducing temperatures during "shoulder periods" or transition times during the late morning and early evenings, which may offer chances for changing the urban design such that more people can spend time outside. If planning agencies are considering options for mediating temperatures to provide pedestrians greater access to outdoor spaces, then reducing the amount of urban area (a direct measure of impervious surfaces) may be the first step. While changing land cover may not be cost-effective or a feasible option in places containing a large amount of impervious surface, making space for trees within highly urbanized areas may be a reasonable alternative. Despite the arid climate, given the abundant amount of water from reusable sources such as air conditioning condensate and treated sewage effluent (TSE) in Doha, water resources for expanding the urban canopy may be readily available.

As expected, we found that the albedo was negatively related to local temperatures in the morning and evening. However, this relationship was positive during midday periods. Although increasing the albedo of surfaces is a common practice, one that could reduce the absorption of solar radiation, it might not work for Doha. Higher albedo could help the temperature in the morning and evening, but it could also have the adverse effect of increasing temperatures during the daytime. Finally, the distance from the coastline indicates that the mediating influence of coastal waters can significantly impact inland air temperatures. If, however, coastal winds are blocked by high-rise buildings along the coastline, inland areas may not benefit from this mediating influence. Restricting development along the coast, especially those buildings that prevent these coastal processes from meditating inland temperatures, is a policy that has traction in scholarly research (Wong et al. 2011), and it may be a policy option that can improve the short- and long-term quality of life of Doha's residents.

Because traffic volume was also found to be a strong predictor of temperatures in major urban corridors, increasing shading measures, such as tree cover, vegetated archways, or other shade structures at key intersections, where traffic volume is highest, and preferably, improving access to public transport to reduce traffic volumes altogether are highly recommended. Currently few cities are considering the explict shading of roadways, though in places where such features do exist, temperature are notably cooler throughout the day.

Again, at the neighborhood scale, the most effective scenario was that of adding mature trees (approximately 10 m height). In the Site 1 case study, the microclimate modeling estimated a maximum hourly air temperature reduction of nearly 1.2 °C and surface temperature reduction up to 13.5 °C at 12 pm. While the scenario of changing the type of pavers resulted in approximately 2 °C lower surface temperatures compared to the base model, we did find that comparing different areas of the base model showed that a standard light red paver will be lower by up to 12 °C than areas covered with asphalt and 3–4 °C cooler than areas of bare soil. This also points to the need to develop improved shading measures for pedestrian pathways and outdoor recreational areas, especially for inland areas as these experience the strongest UHI at midday when both air temperatures and mean radiant temperature are highest.

These studies suggest a need for further and ongoing examination of Doha's urban heat and development patterns, particularly to gain additional insight on the specific planning measures that are most likely to lead to improvements in thermal comfort for residents. Such future research will continue to inform the planning choices, not only for Doha but for the wider GCC region, and help to reduce the possible negative implications of development for humans and the environment upon which they depend.

Climate- and weather-related disasters are not without precedent, and both research and best practices support predictions of potential consequences. In the cases of extreme wildfires and heat events over the past few years, for example, regional authorities provided extensive warnings and substantive scientific and technical knowledge about the timing and extent of these events (Kates et al. 2012). Nevertheless, inefficiencies and failures in knowledge systems—networks of formal and informal actors and organizations where knowledge, ideas, and actionable strategies toward sustainability are coproduced, evaluated, and validated (Miller et al. 2015)— hinder the ability of cities to adapt and reduce the vulnerability of their populations to increasingly frequent extreme climate-induced events. The obduracy of existing infrastructure (Hommels 2005), coupled with entrenched institutional and political dynamics, make cities and their systems resistant to change, thus limiting their adaptive capacity and, ultimately, their resiliency. We argue that only through active and critical engagement between researchers and practitioners to link a dynamic set of knowledge types—scientific, technical, local, and tacit—about interconnected systems will cities advance their climate planning efforts, and indeed adaption practice. Preparation for and response to extreme climate events will become increasingly important as the planet warms and generates greater management uncertainties in safeguarding our cities and regions.

# References

Hommels A (2005) Studying obduracy in the city: toward a productive fusion between technology studies and urban studies. Sci Technol Hum Values 30(3):323–351

Kates RW, Travis WR, Wilbanks TJ (2012) Transformational adaptation when incremental adaptations to climate change are insufficient. PNAS 109(19):7156–7161

Miller C, O'Leary J, Grafy E, Stechel E, Dirks G (2015) Narrative futures and the governance of energy transitions. Futures 70:65–74

Wong MS, Nichol J, Ng E (2011) A study of the "wall effect" caused by proliferation of high-rise buildings using GIS techniques. Landscape and Urban Plan 102:245–253